FLAT EARTH

Valentin Matcas, M.Ed.

Copyright © 2024 Valentin Matcas

All rights reserved.

ISBN: 9781976726071

DEDICATION

I dedicate this book to everyone eager to learn and develop continuously throughout life.

CONTENTS

1 A Delicate Study in a Consensual World   1

2 Real and Consensual Conditions in a Flat World   25

3 Flat or Spherical, Truth or Deception   52

4 Successful Reasoning and Consensual Constraints 62

5 Extraordinary Life in an Extraordinary Reality   68

6 Those in Control of the Human Knowledge   99

7 Our Earth, and the Entire Plane of Existence   117

1 A DELICATE STUDY IN A CONSENSUAL WORLD

Is the Earth flat or spherical? There is only one way to find out the truth. Just go up there in the orbit of Earth, see it for yourself, and then come back to tell us the story. You may always rely on everything that others claim, officially or not, that the Earth is flat or spherical, but when all their statements and proofs remain inconclusive, since they did not go up there in the orbit of Earth to see it themselves, regardless of what they may claim, you have to keep searching.

Yet when you study the current science with all its space missions and space records, even these remain inconclusive, as though space missions have never reached past the lower orbit of Earth themselves. Therefore, now we have additional important questions to ask, besides those regarding the true shape of Earth. Is there anything at all past the lower orbit of Earth? No, according to all space records, regardless of what science may claim.

Because as long as you cannot go past the lower orbit of Earth as an individual or as an entire valid space mission to see everything yourself, and as long as the current science remains incapable to provide the necessary scientific evidence stating that the Earth is spherical and that our world spans past the lower orbit of Earth as we see in the sky, then everyone, the entire humanity remains incapable to prove that Earth is flat or spherical. Since just as all flatearthers and planoterrestrials out there struggle to prove that the Earth is flat, now you have the entire science, with its entire army of millions of employed scientists incapable to prove that Earth is spherical and that there is anything out there past the lower orbit of Earth.

Because this unique circumstance questions not only the accurate shape of Earth, not only the pertinence and reliability of the current science, but it questions the meaning and nature of this entire world. And if you are only starting your search for truth right now, to see if Earth could be flat and not spherical, there is significantly more going on than contorted science and fake space missions, while you should research harder to find out truth.

Because you cannot simply solve a mathematical equation here at the surface of Earth to find out its shape and nature, while you cannot rely on science to tell you the truth as it lies persistently, and now you have to figure out everything on your own.

Yet there are still ways to tell if Earth is spherical of flat, by the laws of physics here at the surface of Earth matching space objects or flat surfaces, or matching rotating objects or stationary ones, as we will see shortly in the book.

Yet you always have to be careful throughout your studies, because many times, created realities are made with the main intention to seem larger and therefore more credible, fooling you the entire time. While dreams and videogames do just the same. And therefore, now you have to consider these details just as well, along with much more.

This book creates a comprehensive model of Earth and of the entire world, studying its shape, boundaries, characteristics, and social circumstances, using reasoning and accurate facts. This research of the true shape and nature of Earth is done from all perspectives: scientific, empiric, social, cognitive, existential, and spiritual.

Because the truth is different than what you are told, officially or not, and now you have to use your own reasoning and not your beliefs in order to find out the truth, complicating everything. Otherwise, you end up with the specific beliefs that others implement and maintain in you, meant now to substitute the actual truth, for various reasons, and always on their behalf.

There are tens of thousands of ideologies in this world, many times contradicting and fighting each other, while the truth is unique, and you know it or you do not. And this is why now, when you seek the truth about the shape, nature, and reality of Earth, you have to find out not only the shape itself, but you have to find out if what you learn is true. While this is significantly harder than you may assume, because you have to use your reasoning, and not laws and beliefs.

Therefore, if you want to know the truth about this world including its shape, you have to know everything about reasoning and beliefs, about truth and ideologies, about empiric, ideological, and analytic research, about the human cognition, human needs, human meaning, and the human condition, about this entire world, and about Life and the actual structure and behavior of the human society, since the human society is built entirely on ideologies of all kind. Because if you cannot distinguish between truth and ideologies, or between reasoning and beliefs, you cannot distinguish between beliefs and accurate facts, you cannot know what is right and what is wrong in this world, and you end up confusing this world with the consensual truth about this world. While these are never the same. And if you persist confusing beliefs with accurate facts, what you learn in this manner becomes what you believe about the real world, and this is not the actual truth. Since this is how you may end up confusing the content of this book with your actual belief about the topic of this book, and again, this is not sufficient to know the truth.

And if you still seek to find out the truth about Earth including its shape, nature, and meaning, then get ready to embark in the study of one of the most controversial topics today, the flat Earth theory. Yes, the flat Earth is back, yet it had always been around, because it had always been accepted in the upper society, it transpires today through them in the media, politics, entertainment, and even in the current science, it becomes more and more visible, and consequently, millions of people already research all knowledge about the flat Earth theory today, very closely. Yet this time it is the other way around, with authorities ignoring the flat Earth theory on purpose, while enforcing only the spherical model of Earth, in every manner, along with the big bang theory, the theory of evolution, the theory of relativity, and the rest of the theories. While just as dictionaries state, the term theory signifies assumption, supposition, or speculation, but not at all accurate fact. With science itself placing at the same level the big bang theory and the theory of evolution with the flat Earth theory, starting with the title. Since theory itself means assumption, speculation, or supposition, but not accurate fact.

But what if you are always eager to accept science and all its valid theories? Then study the current science in all its details, because it is an ideology itself, a scientific ideology. Ideologies are sets of beliefs, and they are based on these sets of beliefs, remaining at the interior of their correlation. The current science is an ideology, since it is based on a consensus, officially called the scientific consensus. Some of the scientific knowledge is based on genuine, accurate facts, as geography, mathematics, and classical physics, but some is not, as most of the modern physics and even astronomy, cosmology, and astrophysics. Making all space missions a theory, now with science lacking conclusive space records. As in the theory of moon landing. This is why most of the current scientific research is ideological in nature, based on consensus but not necessarily on accurate facts, and this includes space science, cosmology, astrophysics, and some of the Earth science. Along with almost everything else, as nutrition, medicine, sociology, biology, economy, politics, psychology, and finance. Since everything remains inconclusive as long as it is kept consensually contorted.

The current science is based on a general consensual agreement formed and maintained by its employed scientist and by those who control them from above, those who control society. This scientific consensus spreads further because it is implemented and controlled by the upper social classes, the current consensual Brotherhood and the Elite, and it is therefore their social and ideological consensus. And this is why you have references at the back of all scientific research today, to confirm their acceptance and conformity to the entire scientific and upper consensus. This is why there are no free of cost energy sources today, this is why you always have to reason on your own, and this is why you have to go up there in space past the orbit of Earth to see the Earth for yourself in order to find out the truth, because no one will tell you anything about the shape and nature of Earth. And if you cannot go up in space, then you have to find a way to reason around these ideologies today. This is rather tedious and sometimes dangerous, and more importantly, you have to reason only through accurate facts, because you have to accept only the truth.

Which ideology in particular should you accept? Which one exactly is right for this research? You decide. But be careful in your decision, not to employ the beliefs of an ideology while assessing the beliefs of another, because this form of ideological thinking keeps you within beliefs, and it happens often. It is difficult to find out the truth, unless you employ your reasoning at its highest capacity. You need an entire analytical, comprehensive model of Earth and of this world to do so, not only an empiric or cognitive model. And what we can tell with certainty from the beginning of this book is that there is something suspicious about the shape and nature of Earth and of this entire world, to have involved this entire human and higher effort now and throughout time. Something is there, but what.

Why can we not use an empiric model in our study of the shape and nature of Earth? Mostly when empiric means observed, and all we need is to go up there in space, observe the Earth, record its shape in pictures and videos, bring them back, and inform the entire world of our discovery, for everyone to know the truth from now on. Certainly, if you can only go up there in space to see Earth for yourself, then problem solved. If not, then it takes more than an empiric study undertaken here at the surface of Earth, in order to consider detail after detail, because you risk considering assumptions and speculations called beliefs instead of truth and accurate facts. And since sets of beliefs form ideologies, you risk ending up with various ideologies, organizations, and societies, now called flat Earth or spherical Earth, which might or might not be true.

Because if these people can go up there in space to see and record the shape and nature of Earth, and if all records and statements remain conclusive, then yes, this is the truth. However, when studied closely, you never have to form beliefs, agreements, and entire ideologies and societies on anything that it is already accurate, real, and the case in this world, because it is already the case. Basketballs are round and this is accurate, while you never find the round basketball society telling you the truth about the round shape of basketballs. They are just round.

And since it is your own ideology this time, and if you have a significant number of followers, now there is a large amount of influence and profit involved, as this is an entire new topic of study altogether. Because throughout your research over the Internet, you have already found the lucky ones profiting consistently from the flat Earth topic at the top of their ideologies, which can be mainstream, scientific, religious, alternative, spiritual, social, and political. Because so many want a piece of the pie for themselves from this topic alone, but not exactly the truth, and this certainly alters your research.

But is the Earth flat or spherical? Which one is true, the flat Earth theory, or the spherical model of Earth? You cannot know it with certainty, just because 'theory' means probable, speculative truth, until proven false. While 'model' means smaller, artificial replica of the real thing, but not the real thing at all. They never call it the model of multiplication, but simply the multiplication table, because two times two always equals four.

Why is the flat Earth theory only a theory? The flat Earth theory remains a theory throughout your study as long as you cannot go up there in space yourself to see Earth in its actual shape, or unless you do not have tangible, accurate evidence of its shape. Or until you cannot correlate the flat Earth theory directly with the accurate facts of this world. And even more, the flat Earth theory remains a theory for as long as the spherical model of Earth remains incapable to offer you sufficient knowledge about the actual shape and nature of Earth. What is relevant to observe here is that the flat Earth theory remains a theory even when the spherical model of Earth fails to prove that Earth is spherical, just because Earth may have other shapes and forms, it may be as large as this world, it may not even be whole, or it may be only simulated.

Does this mean that the spherical model of Earth is true? Is it an accurate fact? No, not at all, or at least not as it is defined by the current science today, just because as stated above, the word 'model' means a smaller replica of the real thing, but not exactly the real thing. While you always use mental models throughout your learning, study, and research in everything related to abstract, subjective topics, as mathematical models or cognitive models, but not with objective ones, not with objects, even space objects. You may keep these plastic models of the real objects anywhere around if you please, yet you always study and learn more about the real things, not about these plastic models. Or you do so only when you cannot reach the actual subjects of your study on your own, but you only assume their form and characteristics in this manner, through your models, since you do not have another choice. In this specific case, science creates this speculative spherical model of Earth, this is what it studies now but not the actual Earth, and so it claims that this specific spherical model of Earth that it studies is spherical. And this is always true, because all spherical models are spherical, as it is trivial knowledge. While the current science is full of these smart, diverting strategies that only seem to be true, but may or may not be true.

Is this an error of judgment, or a diverting strategy? Because we always find these diverting strategies mostly in court, and lawyers are experts by now in manipulating them.

Yet the current science does not use a model of a model as stated above, but it has the multitude of scientific recordings including millions of photos and video recordings taken by the multitude of its space exploration missions allowing it to make now a true model of the actual Earth, exactly as it is seen from space, since it has a multitude of space missions to confirm that the Earth is spherical. Because while you cannot go up there in space to see Earth for yourself, the current science certainly can, and it does so anytime it pleases, through all its missions of space exploration. And now, the multitude of its records that must count by now in millions of pictures, videos, and other relevant recorded data, can prove with no doubt that the Earth is spherical. They teach so even in school, and it is trivial. Because even you take a multitude of pictures of everything that you find relevant in life, you already have thousands or tens of thousands of pictures everywhere about everything, you even have a security camera in your car, house, and at the office recording everything, and consider now the multitude of space recordings proving the trivial fact that the Earth is spherical. You learn it even in school.

Or this is exactly what you assume. Or this is exactly what everybody in the world assumes. That if it is so easy to take pictures and videos using cameras just by pressing a button or clicking a mouse, and with tens of thousands of satellites, rockets, and space shuttles everywhere, now it is very easy to have millions of space records proving trivially that Earth is spherical, exactly as you learn in school, and exactly as you see on TV. But is this accurately true, or only a trick, a lie, a mirage, a charade, or a social stereotype? Just keep researching, to find out the accurate truth for yourself.

Just start immediately to find all these scientific records

yourself, all these millions of pictures and videos of Earth taken from space past and present since they must count in the millions by now, with thousands of live streaming of Earth coming from all space shuttles, space stations, and scientific and commercial satellites counting in tens of thousands. Because recording technology and streaming technology are very advanced, very affordable, and very common today, as they are used everywhere, even in cars, busses, trains, elevators, farms, factories, skyscrapers, restaurants, intersections, and points of interest.

And you can find all these millions of space records to see it for yourself. Just search it and see. You find about nine old photos of Earth taken from space, decades ago. They are of a very low quality, and they just look fake. That is not even Earth, and those are not clouds. Compare now these nine photos with each other, if you can find them, since even these nine old photos are removed systematically from the reach of the public, to see how they are not consistent even with each other. You cannot even identify continents and any geographical detail, so you cannot even tell what you see. While they look fake through many of their details, challenging your intelligence.

Yet there is more. Since in the past few years, the multitude of space missions have added another two space recordings of Earth, one recent picture, and one entire video several months before. And now the entire world is frustrated that while the Earth actually rotates in the video, the clouds remain stationary. Because you cannot fake the movement of all clouds in the world, so they just let them stationary when they pasted them on that picture. While that more recent picture is an actual grey round shapeless, colorless blob, and it could be anything. Because while you can see lies and inconsistency in all details of all space records, why not giving the people space records with no details at all? Yes, just give them pictures of round blobs, to see if they react anymore. And no one actually reacts anymore at this round blob, which is the last picture of Earth from space. No one complains. Yes, it is an accurate blob.

Yet despite these, it is interesting to study how everybody becomes frustrated with the current space records, when the current science and the current society in general lie continuously. Making all astronauts and all employed scientists politicians. And no one is frustrated with politicians, as they always lie. They are politicians.

While the famous blue marble from your textbooks is not an actual space record, despite of what you learn in school. Because the blue marble is only a created, composite image made of thousands of high altitude photos taken from balloons, with readjusted details, added artistic colors, and fake clouds. Just find it and study it in all its legal details, including the fine print. Education may claim that it is a space record of Earth exactly as it is, spherical, while science gives you a fine print. Use this picture only for entertainment. Not for education, not for research, but only for entertainment.

What is going on here? Whatever goes on with Earth, with the people of Earth, and with all knowledge related to these, it remains inconsistent enough to determine the great majority of the most capable people of Earth to embark in the study of this exact topic: the flat Earth or round Earth dilemma.

Then, what is the truth? Because we find these diverting scientific strategies in all research involving highly important and even classified topics, which influence directly the human condition and the human knowledge. Then why exactly is science diverting you, and from what? Because as you notice, the current science informs you in an implicit legal manner that the spherical shape of Earth is not the actual truth or fact about Earth, but it is true only about the model of Earth that it studies, in a consensual, theoretical manner. Otherwise, it was an accurate fact, and not only a model. Because you always find scientific models throughout research every time you cannot perceive objectively the actual object or subject, when you cannot record it directly, and therefore when you have to model it yourself. And judging this specific spherical model of Earth offered by the current science, the spherical Earth is not even a model but a theory, and a rather succinct, inaccurate one.

Now we have to model Earth comprehensively, because we cannot go up there in space to see it in its true shape, and because the knowledge provided by the current science remains insignificant and many times erroneous and misleading on purpose. And it might not even be possible to go out in space far enough to observe the true shape of Earth, for various reasons, the same reasons that the current science also encounters.

But the question always is why this is also the case with the current science, with its outstanding financial and technological abilities. What exactly stops science to go out there in space to record the shape of Earth in all details? Why is the current science calling it a model, the spherical model of Earth? Why not the real thing? Why the big bang theory? Why the theory of evolution? Why only theories and models, but not the real things, the actual truth? Why do these remain scientific theories, which means scientific speculations or assumptions? What is the truth, and where is the actual knowledge about Earth and about this entire world?

Is the Earth actually flat, sitting beneath a dome? This is the belief, and you cannot know it as an accurate fact unless you find the dome, and witness it for yourself. Or if your ideology demands it, now you have to believe in a dome. Yet it does not have to be an actual dome, but only a wall, even a wall of field, or an impenetrable obstacle surrounding Earth, or even an impenetrable outer limit, the dome-like end of this world.

Is this world only as large as Earth itself? Or there is a dome up there, as the Brotherhood and the Elite believe? Then we can never consider the actual shape of Earth, if this world is only as wide as Earth itself, since any comprehensive physical detail of Earth lacks meaning in a world as large as Earth itself, since you lack the outer perspective to consider it, making the Earth flat. Which means that this is the answer, the Earth is flat, if it is a created reality no larger than the lower orbit of Earth, since you lack the means to perceive it otherwise.

Similarly, your videogame worlds are flat by default, regardless if they seek to seem spherical.

What is true and what is false here? You never know it for sure, do you? Or you can know it easily, but only through beliefs, both official and unofficial, and it is insufficient, for any rational, analytical mind. And this is the problem. Because as long as the current science feeds you fake recordings about space beyond the lower orbit of Earth, along with theories or speculations to back them up, then both the spherical and flat models of Earth remain theories, but not facts.

What chance can the flat Earth theory have against the current official spherical model? It depends on who is behind science and how they want the Earth to be, since they can make the Earth flat this decade and spherical the next, if they want, with the entire population cheering the extraordinary achievements of science. But most importantly, it depends on which side of the line you stand this time, because this time it is different, since this time you seem to have a choice in this world, just because you seem determined to conduct your own research to find out the truth, bypassing in this manner the current science and all ideologies claiming to offer you the truth.

It seems to be science against religion and spirituality all over again, only that this time it is different, since as stated, this time you may stand for yourself with your own choice in this world, you may research all facts and premises yourself, you may decide, reason, and accept or reject everything on your own, because this time you are not forced anymore to take sides, to join tight hierarchies and their ideologies, and to drop into servitude. Because this time your place in this world is accounted for, just because this time you are accepted as who you are in this world, as the intelligent, living human being, and not only as what you do for this world, as the disposable assigned role that you play in this world. Yet you always had your choice, only that this time, you are capable to identify and understand your meaning and choice, and this makes the difference.

What should you choose this time, flat or spherical? How should you vote this time? How is it better? This world is more important than a random choice, since this world is extremely complex, with all its laws and characteristics connecting rigidly and therefore determining your human condition in a similarly rigid manner. And it certainly matters, this time and always. And with Earth being possibly flat now and not even spherical, well, this changes everything. Mostly in the beginning of a new age, with Phoenix about to die. Because if science still lies about the shape of the Earth, then it is no telling what it is still true in this world anymore and what is erroneous, hidden, and fake. Because big bang, evolution, creationism, souls, elementary particles, the Deity, scientists, authorities, angels, particle accelerators, and aliens, are related directly to the shape of this world now, to the shape of your own inner world more precisely, and to everything meaningful that you must know about the outside world in order to help you reason at the intelligent human level. And it is very important to distinguish what is true and meaningful from what is fake and irrelevant, just because this type of general, meaningful knowledge creates your cognition from the inside, through your own inner reasoning, remaining always at its base, at the core of your cognitive system. Because you either have accurate facts at the core of your cognitive system to help you reason through them and therefore to help you generate only accurate knowledge and successful ideas, or you have beliefs at the core of your cognitive system, and however you think through your beliefs is not called reasoning anymore, but ideological thinking.

And you may witness right now your strong inner need for accurate, pertinent, important knowledge about yourself, about this world, and about your meaning, place and condition in this world, just because you have to have it at the core of your cognitive system as accurately as possible. Otherwise, your reasoning fails, and throughout life, this can make the difference between wealth and poverty, between success and failure, between freedom and servitude, between intelligence and ignorance, between reproduction and extinction, and therefore between life and death. And it is at this confluence between all your worlds, interior, exterior, and higher, that we hope to find the necessary accurate and higher facts helping us understand the true shape and nature of Earth and of this entire world, the true structure, nature, and meaning of all your worlds, inner and outer, lower and higher, and the true nature and meaning of all your selves, inner and outer, lower and higher. Because the flat Earth theory is not exactly about that brief optical illusion of a flat or round Earth that you witness at the beach while sunbathing and watching the distant ships with a telescope, but it stands at the base of the most important ideologies of all your worlds, inner, outer, and higher, managing in this manner to influence the existence, behavior, and reasoning of all your selves.

Yet there is more to consider, because when anyone is capable enough to distort this specific meaningful knowledge through any ideology, and once you and this world accept it in its distorted, erroneous, fake form, you end up behaving consensually, not on behalf of Life, but on behalf of the multitude of ideologies, or more precisely, on behalf of those who control ideologies, those who control you through them, and therefore those who control this world through you and through everybody else.

What exactly is true and what is inaccurate in this world? There is one simple rule to follow: is it tangible? Then it is objectively real. Do you have it? Then it is there. Did you see it? Then it really happens. Did you witness it yourself? Then it must be true. Is this accurate truth correlated to the base or structure of this world? Then it must be based on accurate facts. It becomes an accurate truth when it is based on the natural laws of this world, otherwise this entire world is not true, it is not real. While all realities are objectively real from their own inner perspectives and as long as you are there, and therefore this world is always real, even objectively true and real, as long as we are here in this world. The existence defining and validating this world, also defines and validates all accurate facts to be true, always true, even by natural laws.

And this is similar with all realities, even with the artificially created realities, and even with the consensual realities. Therefore, is the truth that you consider only a consensus, an agreement among a specific group of people that form an ideology, defining an affirmation true or false within their group? Then it certainly remains true or false, but only as it is defined consensually, and only as a belief to be true or false, but not as an accurate fact. And this remains the case only at the interior of that specific ideology, only among those people, and therefore only within the jurisdiction of that specific ideology. Just because jurisdictions are consensual realities in themselves, they have their own laws, and now, any consensual affirmation or belief within that ideology is actually found at the base or structure of that jurisdiction.

And this is important to consider, because you live your life within various jurisdictions one after another throughout society, you do not live your life in this world as you may assume since they consider you part of their jurisdiction whether you know it or not and whether you want it or not, and therefore you are considered to live your life within these jurisdictions, through your consensual corporation, and not as a genuine living human being. And since all jurisdictions may create and use any law and belief they please, you are expected to live by these consensual laws. It is the same strategy used by the current science to make you accept its scientific knowledge, because the current science is an ideology itself, it has its own consensual scientific jurisdiction, and therefore within this consensual scientific jurisdiction, all its scientific knowledge, laws, and theories are always true, because they are defined there to be true, consensually, by agreement.

Because the multitude of accepted scientists define them to be true, consensually true, just as requested. And now, all accepted scientific knowledge has judicial pertinence and validity in all jurisdictions of society, and this is exactly how you end up condemned and serving time for the model of a model, or for the fake scientific model of a real model believed to have happened and therefore believed to be true. Which does not have to be an actual accurate fact or truth, but only consensually considered true. And many go to jail in this manner, consensually, but they go to a real jail, serving real time.

Or everybody goes to jail in this consensual manner, because the human behavior itself is considered consensually but not accurately to be good or bad, acceptable or unacceptable, social or antisocial. And it is considered to be in this manner in all these jurisdictions through consensual agreements, between a group of people. This is how the few control this world now, and this is how you go to jail. While the human behavior is natural or consensual, but not good or bad. Because everything that you do in life naturally is good for Life, while everything that you do in life consensually, you do so on behalf of those few controlling this world for the Consensual Matrix that controls these people, and that controls this world through these people. And since these people and the Consensual Matrix implement, decide, censor, and monitor the entire human knowledge today and in the past, these decide now how much you should know about Earth and about this entire world.

Because in order to be able to distinguish the truth from the erroneous, you must always be able to distinguish between beliefs and accurate facts, between this world and jurisdictions, between actual truth and ideologies, and between your natural and consensual selves.

While you want the real thing, the actual shape and truth about Earth, and not the consensual beliefs that you find all around about all these. While if you are not careful, you might end up seeking this actual truth among ideologies and within jurisdictions, and it is not there, because ideologies have their own separate sets of beliefs, since all ideologies and all jurisdictions stand apart from the natural living environment, they stand apart from Life and from this world, only not to interact with Life and with this world, on purpose. Because if they ever do, they become part of Life and of this world, and then they have to obey the natural laws, the accurate facts, and the accurate truth of this world, in place of their own beliefs.

They have to tell and apply the accurate truth, and this might get in their way. Because this is why they form ideologies and jurisdictions in the first place, in order to be able to avoid the truth, by defining everything as they please, substituting it in this manner with beliefs, statutes, laws, and theories, as they please.

And this is how all ideologies and jurisdictions define Earth, life, this world, reality, and your own meaning and fulfillment, as everything that they please, by law, by their own law. And it is legal, since they make these laws, in a conflict of interest, always going against you. And now you even learn it in school and see it on TV, exactly as they define it. Because science is an ideology, a jurisdiction, and even a consensual corporation, where it defines by its own laws, beliefs, consensus, and statute everything that you learn in school as a valid fact. They do not call it accurate fact, but only valid fact or scientific fact, knowledge, theory, or law, because they validate it themselves, while it might or might not be accurate, you never know. But they hide all means for you to find out the accurate facts, just as they hide all recordings that they take from space, if they ever take recordings from space, if they ever go in space, and if space itself even exists beyond the lower orbit of Earth. Since now you never know, unless you go there yourself. And there are people who want so much to learn all these, that they even try to go to space on their own, in any manner. But they do not succeed, for various reasons, and many die.

As a reference, all scientific agencies including NASA have their names written in uppercase letters, they are registered as consensual corporations, and they have all the necessary licenses and permits to operate here in the real world. Since they are corporations, and they have their own scientific jurisdictions just as well, to make their own laws and statutes. And this is their consensual environment, not the real world. And this is how they exist consensually, not among the actual accurate facts and not through them, but through the multitude of consensual scientific knowledge that can be sometimes true and sometimes false, since you never know. This is neither right nor wrong, but it means that they may define truth in any manner they please, at the interior of their own consensual scientific jurisdiction. And it is never a lie, it is never fraud and misinformation, and it is always legal, because it is legal by their own laws, in a conflict of interest. And it works. But who are they hiding from?

Therefore, now you know what you want, either the accurate truth, or the consensual information found everywhere, depending on who you are. Because if you are happy with the consensual truth found everywhere, then you already have your answer: the truth is everything that your consensual authorities decide that it is true from one mandate or regime to another. While you can even join them, since the current Lower Brotherhood is open to everybody. And if you are already in the current consensual Brotherhood, then you have to accept the current consensual truth, that Earth is spherical, that there was a big bang, and that life appeared and evolved only on Earth. This is the consensual truth today, and it might even be the accurate truth, you never know, until you figure it out rationally and analytically, on your own. Because you cannot define the shape of Earth consensually, since it is part of the real, material, natural, living world.

While if the current science wants to define consensually the Earth spherical this decade and flat the next, this is exactly what it does, through the multitude of its scientists, all obeying and even recalling now the multitude of anecdotes about them, believing the entire time that the Earth is actually flat, and not spherical.

There is more going on here and we will see it throughout the book, yet basically, the real and the truth are real and true, but you can reach, understand, and use them as long as your mind and reasoning are not clouded and altered on purpose with various beliefs and stereotypes meant to control you and the entire world. However, as you know it well, the great majority of people do not mind being controlled, as long as this helps them keep their job and their place in the hierarchy, in their ideology, political party, hierarchic Brotherhood, or corporation.

Is the Earth flat or spherical? Just let everybody know how to call it this time, and that is exactly how the Earth is. Because as stated above, the truth stands at the confluence of all your worlds, inner and outer, inferior and superior, real and consensual, and you share it through the multitude of your selves, inner and outer, inferior and superior, real and consensual.

Why having to go through this entire bureaucracy of legal diversion, only to find out a scientific fact? What are they after? And can the Earth still be spherical, or it is actually flat? As you already notice, there is more taking place, so be ready to understand everything. Because there is more taking place than the simple social control where they do everything for money or influence, as described by conspiracy theories today. Because there are created realities that are not set in place, they are as dense as they may seem to be, but they may still change, mostly according to people's beliefs. This can still happen with this world, under strict circumstances, yet it can still happen. And when it does happen, your cumulated consensual beliefs, if they are strong enough, may become accurate facts, if the reality itself allows it, through its own natural laws, and through all supreme laws forming and defining it, since this is how you may reach them. One belief that you may share with the entire world can turn now into an accurate fact, while it had always been that way. This might be harmless when it comes out of the mind of an innocent child, yet it can become significant, out of control, or worse, when it comes out of billions of minds sharing the same strong ideological belief simultaneously, and making it happen, billions of minds controlled by a totalitarian Elite, as it has always been the case with Humanity. And if you were ever wondering what exactly is going on in this world, age after age, and it never ends, well, now you know it, since it is always the same manipulation and control not only of the people of this world, but through the people of this world, of their entire reality, timeline, meaning, and therefore fulfillment. Since many times, reality is cognitive in nature, while the consensual can become real just as well, when it can alter the minds of everybody, consensually, through all means, ideological, subliminal, implicit, direct, hierarchical, or judicial.

Does it sound familiar? Can it actually happen with this entire world, even continuously? Can this specific belief be powerful enough to change the weather today, just because it is Sunday and everybody is happy, or just as meteorologists have already predicted? Is this specific belief strong enough to change the entire climate of Earth, just by having the entire world focusing on global warming? And probably avoiding an entire ice age in this manner? Can these strong beliefs of this world change the entire shape of Earth, making it spherical this century and flat the next? Well, science and psychology deny all these, society considers them science fiction, religion promises that Earth is set in place on the firmament just as it is and therefore nothing can be changed anymore, while spirituality claims that humans are capable of anything they want, capable even to change the shape of this world, once they are awaken to their true potential. And with this world developing exponentially while being controlled just as tightly, in a continuous equilibrium, now it is suspicious.

Are these statements true or simple assumptions? Let us check. Science, society, and all those controlling them intend to remain in control endlessly, and therefore the last thing that they want is for anyone strong enough to change the current human timeline in any manner, away from their control. This happens today with the current world order, and it had been the case throughout the ages, with all the old world orders comprising the history of Earth. And this is why witches were killed, for their strong, cumulative will. And this is why you are fed specific medicine, drugs, and food and water additives today, in order to kill your higher mind and therefore in order to kill your higher reasoning and the rest of your higher abilities. This is why your children are diagnosed with autism from early childhood, since they display higher potential even from early childhood, and now, the medicine that you feed them destroys their mind, in order for them not to influence this world in any way, and therefore in order for the entire human timeline to remain stable and in favor of the Elite. This is why many religions promise a world set in place, unchanged, because they make sure through strong beliefs and through strong religious control that this world remains stable in this manner, favoring the Elite along with various higher entities controlling them through the Consensual Matrix. Or this is the case with all false religions, since these false religions do not serve the true Deity, but they serve various being, higher or lower, which only claim to be the true Deity. Or they serve directly the Consensual Matrix, as direct, consensual structures, principles, or concepts.

And so they are served, directly, through the cumulative wish, will, prayers, and veneration coming from all people. To consider spirituality now, by definition, spirituality is meant to place humans in contact with their higher side and therefore with the higher world, and this includes placing humans in control of their entire mind, higher and lower, in control of all their abilities, higher and lower, in control of their timeline, which is the intelligent human timeline, in control of their own life and existence, and therefore in control of their own world, as spherical or as flat as it may be. Yet these are the ideologies of the past ages of Earth, from the time when humans were still in control of their lives and of their world.

And now, you can distinguish the difference and similarity between religion and spirituality, since you may understand them only through humans, only throughout the true meaning and potential that humans have in this world, and only through the intelligent human timeline that should take place in this world. There are tens of thousands of religions in this world, and since they are distinct, different, and many times opposing each other, only one is the true, genuine religion, serving the true, genuine Deity, in the true, genuine manner, through the intelligent human abilities, higher and lower, through the intelligent human meaning in this world, and throughout the intelligent human timeline.

Is the Deity dead or alive? If he is alive, then he is part of Life. The Deity is everything, he is omnipresent, omnipotent, and omniscient, since both religion and spirituality define the Deity in this manner. While cults serve other beings, which may be human or not. Therefore, a living Deity is the entire Life, and since he is omnipresent, he is everywhere in Life, he is all Life, including everything else, ever, at all time. Study Life and the wider world, to find out that everything is alive, not only in the organic form of life, and not only in this world, but everywhere and at all time. What does the Deity want from you? To fulfill your real, natural, living needs, certainly, since through your real, natural, living needs, you fulfill Life, you are part of Life.

What else can you fulfill besides your living, natural, real needs, since by nature, as a living human being, you are real, natural, and alive? You are more, if the people around decide so, since this is the consensual side of this world, and it is not at all part of Life and of the Divine, but it is simply decided by the people of this world, by their own authorities. And this is how these take you away from Life and the Divine, through their entire consensual demands and expectations. Now study all ideologies, hierarchies, and jurisdictions of all types, since these are the ones sending you your consensual needs, to find them diverging consensually many times from your own real, natural, living needs, including your need to know everything about yourself, life, this world, and your place and meaning in life and in this world. And now you do not know these, but you know what others want you to know, consensually. Since this is the difference between the accurate and the consensual.

Why having a multitude of religious and spiritual ideologies in this world? Those who intended and still intend to control humans, humanity, the human timeline, and therefore the entire world, manage to alter or scramble all higher knowledge on their behalf, altering in this manner all meaningful ideologies, altering in this manner the intelligent human development, the intelligent human reality, the intelligent human mind, the intelligent human meaning in this world, and the intelligent human timeline. And they did and still do so in order to exploit one of the highest human cognitive ability, the power of direct manifestation, when cumulative thoughts become accurate facts, when cumulative beliefs become actual events and circumstances, and when cumulative intentions become objective manifestations.

And as stated throughout this book, you are caught right in the middle of these, and once you are made to believe that you have no choice and that you are insignificant in this world, then this is exactly what you manifest for yourself and for your loved ones in this world, that you are powerless and insignificant. And so you become, and so everybody becomes. And this is the difference between the Elite and the rest of this world.

And therefore, this is the kind of world that you create, on behalf of those controlling this world. Yet as stated, this time is different, because this time you have a choice, as long as you are capable to distinguish between accurate facts and beliefs, as long as you are able to distinguish between lower forms of thinking and analytical reasoning, as long as you are determined to develop continuously throughout life, and as long as you are capable to identify and treasure all your abilities, higher and lower. You have a choice in all these, while your choice weights significantly upon this world, so make it a good one.

As you notice, there is more involved in the flat and spherical theories of Earth, and this is why the truth may be significantly further than what you can believe. While we have all the books of this series to help us model everything. And this is why it is important to model the flat and spherical Earth theories comprehensively, consistently, and not randomly or empirically, as many do while they explain separate details of these flat and spherical Earth theories in a separate, inconsistent manner. Because lack of consistency generates only consensual truth, but not accurate truth, and this leads to debates, endlessly. And now, each one of these separate proofs becomes a belief in itself, with the flat Earth theory becoming a genuine ideology serving those at the top, since all truth that you find in this manner is consensual and it remains accurate only within the flat Earth ideology, but nowhere else. While you always want the truth, the actual, accurate truth, the actual shape of Earth as it is seen from up there, from space, if space actually exists. And if you could only get there, you would just witness it yourself.

And as you notice, Earth, along with this world, humans, humanity, life, the human mind, the Deity, religion, spirituality, society, the human development, the human timeline, and the human meaning in this world, are more important understood consistently, through reasoning, accurate facts, and clear correlation, than understood consensually through beliefs, letting them fall into random debates. They have to be modeled and understood through accurate facts and through analytical reasoning, and not through consensual laws and beliefs meant to control your mind even more, along with your behavior, development, choice, and meaning in this world. Because anyone may hijack all these through you, to take control of the entire world. And since it happened in the past repeatedly, it can happen again just as easily. And do not dream to remain at the center of your ideology as you are today, and therefore to get to control this world yourself, because it will never happen, since there are countless of people who want the same, they are significantly crueler than you are and they will dispose of you and your loved ones so fast, despite of what they may promise.

2 REAL AND CONSENSUAL CONDITIONS IN A FLAT WORLD

What exactly is a belief, and what exactly is an accurate fact? Accurate facts refer directly to the natural laws of this world, as the laws of mathematics and classical physics. In general, if you can reason analytically through them, then they are accurate facts. If you cannot reason through them, if you cannot understand them entirely but you have to memorize and recite them wholly, or if they lead to indoctrination or debates, then these are beliefs. Because once you can relate them to the laws of mathematics and classical physics throughout your reasoning, then no one can debate you anymore, and it is an accurate fact. High temperatures burn, or water freezes below zero degrees Celsius, or two plus two equals four, or force equals mass times acceleration. However, throughout many created realities, you may change some of their natural laws through the direct power of your higher mind there, causing many beliefs to become true and real then, and this is the case mostly with the specific beliefs that are allowed to influence and change that reality. While two plus two will still equal four, yet it always depends on the specific reality where it takes place, and on your abilities. And this happens many times throughout your dreams and projections, yet it can still happen even here in this world, with or without your knowledge. And this is how you may end up changing your own timeline through a simple wish, prayer, or belief, to make it better or worse, and you might never know.

And this empowers the Elite today to control this world, since the Elite controls this world through you, regardless if you are from the Masses or the Brotherhood, through you allowing and empowering the Elite to control you, through you controlling yourself and those around on behalf of those above, and by you manifesting directly the strong control that the Elite has upon you.

What exactly does the Elite do in this world? Nothing, in order not to take responsibility for it, but they only profit from you and from your higher abilities, since you manifest for the Elite everything that they ever dream of having and exploiting, including you and the entire world. Yet there is more to consider about the Elite, since the Elite is part of an entire system of consensual existence and control, the Consensual Matrix. The Consensual Matrix operates here on Earth with all the necessary permits and licenses. While all jurisdictions here are part of the Consensual Matrix, with the entire bureaucracy of Earth meant to remain consistent with the Consensual Matrix, and through it, with all life in the wider world.

As you may notice, this world is very dense and very consistent, meant to hold its natural laws and continuum accurate under all circumstances. Or this is the official statement, and probably this is why witches were killed, in order for this world to remain dense and stable, without their direct influence. Yet if you manage to control everyone in this world, if you manage to implement consistent beliefs in everyone in this world, you manage to bend reality itself, up to the point when you are capable to create your own, accurate, consistent timeline, as different from the timeline of our Creator. And this is exactly how co-created worlds become flat, spherical, globular, convex, concave, linear, digital, cognitive, or consensual.

And this happens in this world today, and this is how the Elite manages to subdue, control, and profit of the entire world.

For now, from your own perspective, as a reference, your laptop or tablet are objectively real, along with everything objective surrounding you, along with everyone around, along with everything that you have ever experienced yourself, along with everything that your credible and trusted sources claim to be true. These are accurate, objectively accurate, or at least they are accurate for you. Yet in what it concerns the nature, meaning, and shape of this world, well, these are hard to determine, just because no one seems to be able to reach the comprehensive perspective of the entire Earth in order to perceive the entire world at once, to witness, reason, understand, and therefore decide for themselves the truth from the false, the accurate from the erroneous, the genuine from the fake, the good from the bad, and the meaningful from the irrelevant. While this incertitude leaves room for lies, mischief, and discrimination, and through these, this is exactly how you are determined to accept your control by strong authorities meant to harm and exploit you even indefinitely. Since just because anyone or anything may hide themselves under strong concepts and authorities such as science, religion, spirituality, justice, virtue, democracy, and even Life and the Divine, just because anyone or anything up there may impersonate any of these, this does not mean that they should get your will, cooperation, and vote in this world this time, as they always do. Because you have your force and choice in this world, yet you have your responsibility, and along with your nature and achievements in this world, these four define your status and rights in this world, determining the force and weight that your choice has in this world. Neglect only one of them, your responsibility, and your force and choice in this world may end up serving someone or something else, allowing them to harm and exploit this world through you, again, endlessly. And if it happens again as it always does throughout the ages, then this fake knowledge about Earth deliberately implemented in this world is doomed to return now and every time in the future, decades, centuries, millennia, and ages down the road, to haunt you, your descendants, and their descendants, until someone gets it right. Until you manage to understand this world not one shape at a time regardless of how distinct these seem to be from the ground level of Earth, not one characteristic, nature, meaning, or law at a time since these count by the zillions, not one authority at a time since these wait in line to exploit you, but you can break free only if you manage to understand this world wholly, everything at once, through accurate facts and not through stereotypes and beliefs, understanding it in a rigorous, analytical, comprehensive manner, once and for all.

Well, but what model of this world should you choose this time, and what exactly should you do? Even more, how exactly can the Earth be flat all over again, as people thought back in the dark ages, when they rode horses and hunted witches? Yet people blamed people of witchcraft back then in order to get them out of the way, just as people blame people today of a specific sexual activity, also to get them out of the way. And then you see it in the news, since it tends to happen with politicians and celebrities, for their influence and money. Because nothing ever changes in this world, and these same people punish you now for claiming that the Earth is flat, just as they punished you for claiming that the Earth was spherical in the past. Because, through the use of your own reasoning, it is science that you are opposing today, while opposing your current political party and the current rulers of your nation, along with your current ideologies, as hidden as they may be, and along with the entire world, since your authorities are capable to summon the entire world against you, if you only diverge.

And now, depending on the true nature and shape of this world, these authorities claiming to own this world might not be human, and might not even be of this world altogether. And this is why your reasoning counts now, allowing you to be able to identify and understand this world, these authorities, these beings, and what these beings do to this world. Because your force and your choice matter in this world, in order to allow the entire world to interact with these beings in a most adequate manner. And this is why you count the most, since you can empower this world to deal with this matter, once and for all.

Yet Earth would have to be either flat or spherical, wouldn't it? Besides, how exactly could Earth be flat? Would this not move continents around and everyone would notice it at once? What about all flight delays in all airports? If this was the only issue with the flat Earth, because as you notice, your current knowledge about Earth affects the faith of many, just as it affected the many in the past whenever people failed to cope with the tight requirements of their religious, national, and civic authorities.

Because you are highly significant in this world, since everything is connected, with you being caught right in the middle wherever you are positioned in society. And it is this specific interconnectivity that you should account for and study for yourself, but make sure that you use your analytical reasoning and not your lower level thinking and beliefs, since you are a human being, and you are capable of strong, rigorous reasoning. And probably this is why Life brought you here as a human being, to reason, understand this world, identify problems, find solutions, and save this world. Yet even through your intelligent human cognitive abilities, you are in for a tedious research while understanding this world, since you have no premises to start with in order to understand the true shape, nature, and meaning of Earth, since science and education teach you one model of this world, the globular one, the one that the entire world shares today.

Try to study Earth in any other manner and perspective, and you already have your answer right now and you knew it since elementary school: the Earth is spherical, 4.5 billion years old, placed in the Solar System, with the Solar System going around the galactic center, as it experiences global warming and poverty issues.

And what if you have actually implemented this knowledge

yourself alongside everybody else, including all poverty, misery, loss, austerity, control, harassment, and exploitation in this world? And if you have created all these, then what exactly are they making you create this time, also on their behalf? Can it be an entire flat Earth, where the Elite may step out in the open to be venerated as true divinities by the entire world, again, as it had always happened throughout the ages? And aren't some people already striving the most to create this new flat world, while they are only trying to prove another conspiracy theory in this world, the flat Earth model, even if they do so empirically, one simple effect at a time?

And as I state it continuously throughout this book series, once you distinguish, identify, and understand beliefs for what they truly are, once you base your reasoning on accurate facts, once you avoid drugs, divertissement, dogma, ideologies, and social control, and once you are determined to develop continuously throughout life, you can make things right.

Can this world be flat, square, spherical, or in any other shape and figure? You will never know it, since you can never get high enough above Earth to see it for yourself as it genuinely is. Yet it is possible that the Earth is of many shapes and forms, depending exactly on what people believe in one age or another. This seems impossible to happen and yet it may, you do so yourself, knowingly or unknowingly, and as you can see, everything relates with your creative nature, and this is exactly why your choice, knowledge, and beliefs matter in this world.

Luckily, there is the Internet to help you decide, since it is full of pictures of Earth as seen from space wholly, the beautiful blue sphere that you have always known. Study these pictures now to find them artificially constructed and beautifully composited in every manner, to make you admire them and purchase them. Because the Internet is more of a consensual marketing environment than the useful learning and social environment that it once promised to be.

Now search not for wallpapers, but for the genuine pictures of Earth, taken from space, taken during all space missions or taken by all satellites placed in the higher orbit of Earth, and you find none. Or you may find even more constructed images created artificially here on Earth according to people's beliefs, and they can mislead you. There are tens of thousands of satellites in the higher orbit of Earth, and none has a camera, none streams live the image of Earth as seen from above. Out of the multitude of space missions, there is only one picture of Earth to have been taken from space, in 1972. Yet there are eight more 'genuine' pictures of Earth taken from space, and if you study them closely, you find them inconsistent even with each other. Just find them and study them to see the fakery for yourself. And now, all genuine images that you may find are made on top of these ones, as composites, or they are constructed, while having these specific pictures as a main reference. Study them to find clouds placed randomly, even cloned, without respecting the laws of physics on Earth. Or having continents placed at wrong scales. While this specific 1972 picture is a composite of a multitude of pictures taken from balloons as science claims, but it is the genuine picture of Earth as seen from space as science also claims, so nothing really makes sense.

Or everything is part of a greater strategy, a greater scheme to make you believe in a flat Earth now, and through your controlled beliefs, to make you create not only a new flat Earth, but to make you create the entire ideological stage necessary for all old beings to return and rule Earth in the open once again, as they did throughout the ages, and as they do now from the hiding.

Where exactly are the pictures of Earth taken from space by all satellites? Since satellites count in tens of thousands by now, and they have cameras. Maybe that you have not searched the Internet hard enough. Or maybe that these nine photos are the only actual, genuine pictures of Earth taken from space, as altered and as fake as they may seem to be, probably because Earth itself is fake and altered, while you never expect. A better question is where are the artificial satellites of Earth, counting in tens of thousands by now, because when you look up in the sky at night, you see only airplanes, meteors, and very few occasional satellites orbiting Earth on lower orbits, as they did since the seventies and long before. While these might be natural satellites of Earth, not artificial. They never send satellites up there. Who needs technology in another dark age? Or who knows, there might not be space up there at all, with this world spanning only as far as the lower orbit of Earth.

What exactly is going on here, and why is no one aware of any of these? Probably because everybody is more interested in entertainment and in how presidents tweet and interact with each other. Or probably because everybody is more interested in who killed what president last century, why, and how. What else should there be more interesting? The entire world. Because whatever goes on in this world, you have to understand it as a whole and not one detail at a time, because details count by the zillions, and they are meant to keep you distracted. But to answer the question, what goes on in this world exactly is that no one goes up there in the upper orbit of Earth and beyond, not you, not those around, not your nation, and not even the mighty science, with all its space agencies, no one goes up there.

And what is more important, no one cares. And once you do not go high enough above Earth, you cannot see Earth wholly, and therefore you cannot see its shape. And what science can see up there from as high as its scientific balloons can get is a flat Earth, which could be the actual shape of Earth, or it could be only a matter of perspectives of an actual spherical planet, since you cannot tell. The word 'planet' itself means part of a plane, or a smaller plane. While flatness is only the characteristic of a plane.

Yet the Earth does not have to be entire at all, but only its specific part where you live may exist, your surroundings. While the information that you receive about the rest of the Earth is only fed to you systematically, only to give you a larger impression of an entire world. And even more, the Earth does not have to be real at all, with all its details, characteristics, and events to be only fed to your senses of perception directly, by an extraordinary processor, making you believe the entire time that everything is true. Notice how, in all these examples, the shape of Earth is irrelevant, as long as the Earth is not entire, and as long as it is not real.

And this is how you never get to know the true shape of Earth, but only whatever they decide to inform you about Earth. While you cannot go up there in space to see it for yourself, and while science itself, along with everybody else, cannot offer you relevant evidence that the Earth is as they claim to be. Probably because accurate knowledge itself is limited in this world, spreading only as far as this world takes place, at the surface. And now we have to research these new premises throughout the model. How many more?

There is one important detail to consider. If this world extends past the upper orbit of Earth, as far as we may see in the sky at night, and if everything that we see in the sky at night is real, including all stars and even all galaxies, and if life spans there, intelligent, civilized life, then their laws synchronizing their common existence may forbid them to come in contact with any primitive civilization found below the stage of space exploration. In other words, as long as the people of Earth remain below the higher orbit of Earth, they are considered primitive, and no one is allowed to interact with them. And if you see how space shuttles do not go past a specific altitude but assume a horizontal trajectory below it, probably this is why. Even more, the Consensual Matrix might not be allowed to operate on Earth and exploit the people of Earth as it does, if they are developed past the stage of space exploration. And for this reason, the Consensual Matrix is determined to maintain the entire human civilization below its space exploration stage, indefinitely. And if the people of Earth start believing in a flat Earth, it is even easier. And now it makes sure that only obvious fake records of space exploration exist on Earth. However, all these are simple suppositions, and as long as we do not have tangible evidence that they are true, we cannot include them in our study.

This is the problem, and this is why you are never shown

pictures of Earth but only constructed images, chunk by chunk, as a great game of puzzle, all components taken from as high up in the air as balloons can reach. While the upper society calls that upper limit of Earth a dome, as in a glass dome or a force field, keeping everything confined. And what is significant, this matches religion closely. And not only this detail matches religion closely, but there is an entire movement taking place in this world at a very large scale, attempting to reenact religious events very closely, and therefore attempting to bring back those old timelines as close as possible to how they were back then. The dark ages. Existence tends to manifest through and within distinct timelines, and you have to adopt them in your attempts to implement your own timelines, otherwise you risk failing. I refer to these as existential eigenstates. Because probably someone or something is trying to have a private, venerating planet instead of a private, enslaved planet, and there is a great difference between the two. Or this world is more as a prison, and therefore they make sure that none of the available scientific records shows people exiting the lower orbit of Earth. And suppositions are many to make. Because all space records are so obviously fake, that all movie studios are by far more capable to make better fakes. Space records taken by Chinese space exploration include even burning meteors as they fall down in the atmosphere of the Moon, shadows that are not parallel, and sand displaced by water hoses.

Why do people fail to understand this world for what it truly is, and for what it truly intends? Because everybody does the same mistake, including science and all scientific authorities. Beliefs are used by all higher social classes including the Elite in their reasoning, leading them within the same ideology. Or they try to elucidate this mystery of our world by attempting to understand one detail at a time, many times in an empiric manner, while they are supposed to understand Earth and this entire world analytically and as a whole, since everything is connected. Because as it seems, the entire world might be no larger than the surface of Earth itself, and this is why Earth looks flat, because all surfaces are always flat when considered from within. This is why there is a dome up there, because there is nothing past it and the limit of our world looks and feels as a dome, and this is why all the lights that you see in the sky resemble to an entire universe, which may be real or not.

But where are the genuine pictures of Earth taken during all space missions, hundreds of space missions, public and commercial? Just keep on searching the Net and who knows, maybe you manage to find them, including the pictures of Earth that all astronauts took from the Moon during all lunar missions. Because you cannot find them anymore, since they are removed systematically, whenever people find inconsistencies in them.

Why now? What happens today is that the entire world takes pictures, many pictures. Everybody uses image editors intensively, and now the entire world is capable to spot fake, composite images. And this is why all pictures and videos of Earth from space once claimed to be genuine are taken off the public records and are discarded in batches, hoping that no one remembers them. This is what the right to be forgotten is all about, since fakery happens in all domains.

And now, without any of these pictures taken from space, you cannot prove that Earth is flat or spherical. What else happened then? Who did everything? How exactly have you contributed? What can you do not to let it happen again? And now, how exactly can the actual shape of Earth ever help?

Is Earth flat or round? Yes for both, since disks are flat and round. But is Earth flat or spherical? You can never know, since you cannot go up there to see it for yourself. And now, by seeing all fake pictures of Earth posing as genuine material, you cannot trust science either. Or you have to trust science, if you happen to be a scientist employed by science and academia. But now, since you cannot tell if Earth is flat or round, you have to join in this new debate alongside everybody else, in order to make your point. Or this is the case if you happen to think through beliefs, because if you employ analytical reasoning throughout life, you will engage in research and not in debates, using accurate knowledge and not beliefs and stereotypes, because only beliefs and stereotypes lead to debates. And since there is no accurate knowledge for you to use, by lacking the actual records and all evidence, what do you do?

And what is more important to understand, through your beliefs, you are capable to engage science in any debate, at any time, and even win, because science is built on fake knowledge, beliefs, and consensus, and is not capable to keep its lies consistent anymore. Because this world is very low and greedy today, with no one capable to keep track anymore of why it is in this manner, how it got here, and on whose behalf everything takes place. Yet you can understand everything if you are capable to use your accurate knowledge and your intelligent human reasoning.

Would it ever make a difference if Earth is flat or a little curved? Everything depends on who you are and on what you expect from this world. For some people it makes no difference, and they are ready to claim anything, as long as they can still watch cable and walk to the fridge to get another beer. Others care for the truth, and every time they stumble upon unusual discrepancies in this world as this one regarding hidden characteristics of Earth, they always take the time to investigate and find the truth. Yet in this case, the truth is so inaccessible, while your research is so minutely diverted from all consistent results, that it does not really make a difference if you are a skeptic, a scientist, a strong believer, a spiritual being, or a genuine flatearther, because this time there is nothing to help you, unless you are capable to engage in a very tedious research starting from scratch, involving not only the entire world, but yourself, your reasoning, your needs and meaning in this world, along with Life, society, and the entire human civilization, along with their meaning in this world. Because there is more involved in this study, since the most powerful stereotypes in this world are there to control your thinking through the most common beliefs, with the truth remaining far away from you. And to make matters worse, the closer you get to the truth, if you ever do, then the more your credibility in society is altered, on purpose, so no one will ever care to learn anything from you, if you ever succeed throughout your research. And this is the exact circumstance with the flatearthers and planoterrestrialists when they manage to succeed throughout their research. And by the way they are undermined by society throughout their mainstream research, we can tell that there is something going on with this specific topic, the flat Earth.

As stated above, if you start this research using all current mainstream scientific knowledge, you get nowhere, since you remain within the current scientific knowledge, because science confines your thinking within predefined pathways, through systematically implemented beliefs and stereotypes targeting you in this manner since childhood. Therefore, if you now find globes in all classrooms, if you find pictures of the spherical Earth in all news and movie, if you find drugs and entertainment all around, or if you find yourself being minimized, made to seem and feel insignificant, meaningless, and worthless through all lessons at school and through all TV channels including Discovery, this is exactly how your mind is clouded and disabled, to stop you from reasoning for yourself, to determine you to seek pleasure and material achievements, and to accept hierarchies and authorities in order to do so, to drop into servitude, and therefore to be exploited and to exploit others alike. Because no one exploits the people of Earth, at least not officially, because the people of Earth always exploit themselves, and they do so through their own greed, ignorance, and inability. Or this is the official version, and now this is what everyone out there assumes about Earth.

Therefore, if you ever engage in an independent research of the flat and spherical Earth theories as you see everyone doing on the Internet, this is exactly how you end up modeling the force of gravity in a flat world using density difference and continuous dynamic acceleration, this is how you end up modeling the force of Coriolis using chance, and this is how you end up modeling the edges of the flat Earth using actual domes. Because every time you try to step outside science in order to investigate anything on your own, you never succeed, or you never succeed by using the current laws of science, since some of these laws and theories are not based on accurate knowledge as you may assume, but only on consensual beliefs, shared by all employed scientists of this world. While these beliefs are fabricated in this manner in order to fit all current models implemented by science: spherical space objects, global warming, big bang, relativity, and evolution.

And this is the case with all beliefs, since every time you employ beliefs and stereotypes throughout your independent research, they take you nowhere. And this is why you have to base your understanding of this world on accurate facts, which you may understand yourself well, and not to base your understanding on beliefs and stereotypes. Yet in order to understand this world in an analytical, rigorous manner, you have to take the time to reason systematically throughout life, in all circumstances. While this already seems to be such a ridiculously improper activity to perform. Do so for delicate topics as the flat Earth theory, and you are already labeled as abnormal and antisocial, since this world is more eager to take drugs and to be entertained. This is not abnormal at all, but this is what you are expected to do. And since society offers you no accurate knowledge to reason throughout life, while it offers you an abundance of drugs and entertainment, here is your vicious cycle, keeping you engaged in lower level activities for life, alongside the rest of this world. And therefore, this is the kind of Earth that you now create, full of dualities.

Would you like to employ accurate facts throughout your reasoning and not beliefs? Then you have to do so, by using knowledge involving accurate facts and not beliefs, yet you can find none in science. As a reference, all accurate facts relate to classical physics, electrodynamics, and mathematics, with the rest being based on beliefs, as the theory of relativity, quantum mechanics, and the rest of modern physics. And this is how science remains similar to religion and spirituality today, through its lower level structure and purpose in this world. You may still find some genuine science, religion, and spirituality in this world, always on higher social levels. Or you may find them even on all social levels but very rarely, among people who are capable to reason for themselves and understand this world on their own.

The Earth could be spherical, but how exactly can you tell that this is accurate knowledge and not another belief propagated by all social domains including science? How exactly can you find out? Yet every problem has a solution, and therefore you can tell easily if Earth is flat or spherical, even from the ground level, in a multitude of ways.

Then where is all accurate knowledge if all we get are beliefs and stereotypes even throughout education? Everything that we get from those around are beliefs and stereotypes, if we are not careful and assimilate everything wholly, and if we do not reason continuously in order to judge for ourselves the validity of everything that we learn from them. Including the most common knowledge, since these are filled with beliefs, and we must break them down to their smallest cognitive elements in order to understand them for what they truly are.

Common examples of beliefs and stereotypes are that your own nation is the best in this world, or that your Brotherhood is superior to the Masses, or that everything that you know about this world is true, or that all your attitudes are relevant, or that only your own religion and spirituality are genuine, or that everything that you know is always true.

Is this not your case? Are you ready to accept that what you already know and think may be erroneous sometimes, and therefore you are always ready to learn and change your beliefs with ease? Just make sure that you change your beliefs with accurate facts. Since all accurate facts are connected, and they connect with the natural laws of this universe, as the laws of classical physics and mathematics. In fact, mathematics may be applied throughout other realities, or throughout most of them, while classical physics applies only to this world, as it applies throughout all naturally related realities linked vertically with this one, higher and lower. You may relate any accurate fact with the classical physics, including social, biological, geological, and psychological facts. And once you do, you are capable to use analytical reasoning throughout all domains of this world, as we have seen throughout all models of this book series.

Yet many times, you are placed within loops of reasoning from the start, even from birth, since this is what it is assigned to you through your human condition. You can never understand our higher reality or realities, and you cannot even assume that these exist, even by using analytical reasoning.

Similarly, you cannot know the true shape of Earth, since you cannot leave Earth, and you cannot go into space to see it from there as it truly is. There is a curvature involved, since distant objects across water seem to submerge with distance. And therefore, very far in the distance, you cannot see boats anymore, because they drop under the horizon or under the water level, because Earth is curved. This is the case, and therefore Earth must be curved, right? Now use the distance to these objects and their height, and with simple math, you may find the radius of Earth, concluding in this manner that Earth is a genuine sphere. And this is the case just because light travels in a straight line. Yet if you use a pair of binoculars, miraculously, you are able to see the boat again, even from below the curvature of Earth. You may conclude now either that Earth is not curved but flat, or that light travels in curved lines to parallel the curved surface of Earth, which may be possible during exceptional optical circumstances when the air right above water varies its coefficient of refraction, keeping light always parallel to the surface of Earth, as flat or as curved as this may be. Or this is what science and society claim every time flatearthers try to prove that Earth is flat, and this fuels even more these debates.

Because you need very specific conditions in order to keep the rays of light systematically curved in order to match the curved surface of Earth, mostly if there are larger waves present. While light always curves itself following the curved surface of Earth, under all circumstances, or this is what you always see. But then, even with your pair of binoculars, you may watch that boat submerging slowly under the horizon, as it speeds away, proving once again that the Earth is curved. But bring even more powerful binoculars now, to be able to see the boat again right from under the curvature of Earth. Because Earth is flat. But wait fifteen minutes or so, to see the boat disappearing under the horizon again, so Earth is a sphere. Just use a powerful telescope now, to see the boat again, from far, far away, to prove that Earth is flat after all, and then to watch it submerge under the horizon again, as it speeds away, since Earth is a globe.

What exactly is going on here? How exactly can you pull a boat up from under the curvature of Earth to see it again? Note that this entire time, the observed boat does not dissipate in the mist or polluted air, but it genuinely submerges under water, as though Earth is genuinely spherical.

Why the duality? Can Earth be round and flat simultaneously? Science never approaches this experiment, in order to stay safe. Or it simply claims to be a mirage, while flatearthers use it to prove that Earth is flat. For flatearthers, boats cannot be seen anymore simply through lack of perspective, and not because they truly drop under the curvature of Earth. Yet flatearthers never attempt to understand this specific vanishing perspective either.

Is it a duality between science and flatearthers taking place? Notice how science remains incapable to model its most important achievement ever, its globular Earth model, having to stand now at the same level with any flatearther out there throughout all their debates. And there will be a large amount of these debates coming in the near future, so stay tuned.

There are many experiments to state here, all leading to the same results: the Earth is either flat or spherical, its radius is well determined or not present at all, Earth is placed at the center of this world or on a specific orbit of the Sun, together they circle the galaxy or they go nowhere, while there are similar galaxies in the universe counting by the trillions, or none actually exist. We end up finding similar results with science or we end up proving that Earth is actually flat and at the center of the universe, placed there by the Creator and set in place indefinitely, with all visible space objects circling once a day.

I can state these now because my models end up to be more consistent with religion, spirituality, and with all testimonies of those who left this world and then came back to tell the story, while I can still apply the beliefs of science to prove otherwise, that Earth is globular all along. Again, we have to stumble throughout this duality now, at the beginning of our model, only because later on, we may understand what causes them. And everything relates with you being placed on the same lifeline of causality with all these, influencing them directly, through higher powers, alongside the rest of this world, as demanded by the higher beings above, as they are caught in a similar higher experience.

Yet this is common cognitive procedure throughout all cognitive realities, since comprehensive reasoning is possible only through the reenactment of higher worlds taking place within inner, cognitive, subjective realities. Which means that, however the Earth is throughout the higher realities that it reenacts, this is how we have it here now. Yet Earth does not have to be part of its surroundings, but only of the specific scenery significant to this specific reenacted higher cognitive procedure. And now, with Earth in a world that is just as wide as it is, or even smaller, its shape here in this world has no meaning, since there is no outside, and therefore no space to help you place your perspective and witness it accurately, making the Earth seem flat. But once they need more than Earth in the cognitive reenactment from the higher world, Earth becomes spherical, just as all its terrestrial laws and effects already depict.

Who exactly can leave this world and then come back? You do, at birth, at death, and every time you manage to project. You leave not only Earth, but also this world, entirely, and then you come back to tell the story, if you ever decide to do so, and if you know how to come back. And the story that you always tell is the same, that Earth is plane. Or more precisely, that Earth is a plane of existence in itself.

While it is characteristic for religions to change words on purpose throughout time, in order to change their meanings, and this is how you end up with the word 'flat' characterizing Earth today, and not the word 'plane,' as in a plane of existence. Similarly, higher beings, while coming down in this world, always come into the daughters of men, but not directly into men, for various reasons, and never into the wives of men, for various other reasons. Yet the entire expression of coming into someone of a lower existential level is misleading in this manner, otherwise, you managed to understand Earth for what it truly is.

Because once you understand Earth as another plane of existence, the dilemma of its form and shape becomes meaningless. Because as stated, with Earth being this world itself, its outside shape has no meaning, since there is no outside for any reality, while you can never witness the shape of an entire reality.

Yet these testimonies are beliefs, along with a multitude of religious and spiritual beliefs, while the task of this entire study is not to repeat scientific, religious, and spiritual beliefs, since anyone may do so. But our main purpose is to understand Earth and this entire world accurately, as science was supposed to explain it to us, if science was genuine.

What exactly is wrong with science now? It is based on beliefs, and it performs empirical studies and not genuinely analytical ones. In fact, you cannot use beliefs and inaccurate knowledge through analytical reasoning, but only through lower level thinking. A common example of erroneous reasoning and therefore erroneous research is to observe effects and then to explain their causes in any credible manner that makes sense to you and to this world, without finding their accurate causes. The official purpose of science as it claims itself to be is to fulfill people's need for knowledge. If this may seem relevant to you, then you are exactly what society expects you to be. Yet many people expect science to offer genuine knowledge, as pertinent as possible, since their lives depend on it many times. And this is why science changes its explanations continuously, in order to match people's continuous change in their needs for knowledge, just as it always states that it should do. And it is a characteristic of all domains of society to state their tendencies and intentions from the start, and then to do exactly as they promise. Yet many people cannot understand what they state initially and why they do so, and then they do not understand why all consequences tend to be against them, never in their favor. Because everything is stated in the fine print, while you agree with everything, ending up harming and exploiting this world on behalf of all those controlling it.

And with science modeling effects, it can never find, state, and explain the main causes generating these effects, and therefore it can never model the entire circumstance, but only its visible parts, its effects. Yet flatearthers do the same, since they also attempt to tackle all problems individually, and they never actually model the entire flat Earth theory from all perspectives, as we attempt to do throughout this book. And it happens the same in medicine, since medicine always makes you feel better through drugs, but it never cures you of the actual illness. This is why medicine calls its medication drugs, because this is what they are. And therefore, you know it before you take them, to feel only better but not to get better, not to be cured. Because the actual illness causes your symptoms, and now, by not curing the illness but only its effects, you are doomed to remain sick and to come back for treatment indefinitely. While this is called terminal illness, and everybody dies in this manner.

Because you must consider main causes, and you have to model these just as well, even before you model their effects. Or if you do not do so, you risk falling into misleading loops of reasoning, giving you erroneous results.

Anyone can find the redshift of distant galaxies, just by

measuring it through common spectroscopy. Yet by assuming that the main cause of this measured redshift is the expansion of the entire universe, well, through what premises exactly do you assume so? Because the truth itself may come in all forms and shapes when you perform your research backwards, empirically, from observed effects toward assumed main causes.

As a reference, through the relativity of our spacetime continuum, the entire universe does not have to expand, as long as time slows down with large distances at the scale of the entire universe. And with a static universe now, well, you can never have big bangs anywhere in the universe.

And what if the main causes themselves are not even there, but only their effects? You would never know it, would you? In videogame worlds, nothing is real, not even the entire Earth, but only the local surface is present, only the ground is there, for game characters to walk on. While gravitation is always present even though the entire Earth is not there in the game to cause it, since it would be impossible for the entire Earth to be there. Only the law of gravitation is present throughout the videogame, holding all characters down to Earth directly, through computer code, while they may interact accordingly, and seem realistic.

Gravitation is simply there not because of the entire Earth from the videogame world pulling down objectively on all objects and subjects, but the gravitational force is present in that videogame world through assignment, because it is embedded within the equations of motion of all objects and subjects of that videogame world. Because programmers use the formulas defining the real gravitation in all equations of motion from the videogame, in order to make all motion within the videogame as realistic as possible.

And this is the case with all artificially created realities, because the main causes do not have to be there, but only the laws and effects that they generate. Similarly, in beautiful, exotic settings of the videogame, the sun is not actually real, but it is only a set of layers of images and colors on the display.

While its light and heat are real in the videogame, and now your videogame characters tend to be more determined to go in the water to swim and play.

And if this world is a created reality, even a naturally created reality as these tend to manifest throughout ordinary and extraordinary cognitive systems, then this is exactly what we have to consider and distinguish throughout our research, what is genuinely real and what is only assigned or assumed real in this world, by default. Are the Sun and the Moon there as genuine, real space objects in this world? Or they are only simple assigned images in the sky, while only their effects affect Earth and the people of Earth now, by default, directly through the natural laws of this world, as heat, gravitation, light, tides, solar flares, and orbital alterations and synchronizations within an entire solar system.

And it is the same with all space objects that we may see at night. Are they objectively real in the sky, or they are only assigned, assumed, or expected images on the firmament of Earth, rotating around Earth once a day along with the Sun and the Moon? Because the more our understanding of this world advances, and the more our videogames advance in interface, artificial intelligence, and perfection of details, and the more religion and spirituality become strikingly consistent as we compare our world with our videogame worlds. And this is how you may ask now if the Earth itself is genuinely and objectively real here in this world, or only its surface is genuinely real, to constitute now your solid ground and to give you your assigned gravitational attraction, along with its curvature and rigidity if needed, and along with its shadow on the Moon during each eclipse of the Moon, but nothing more. And are you not caught right in the middle of these manifestations again? Since you can create everything regarding Earth and the effects that Earth has upon this world, from rigidity to its attraction and curvature.

And this creates all dualities in this world, because throughout specific ages of Earth, people considered, expected, and created Earth with a flat surface. While during this age, you attempt to create everything round, on top of what it is already flat, rigid, and definitely set in place.

Yet there is more to consider, since throughout this lifetime called Life on Earth, you are taken from one chapter to another depending on your success, learning, and achievements. While all chapters remain consistent in general, so you never know that you are taken from one chapter to another, and therefore from one actual reality to another, all called Earth. Yet you may still find minute differences at times, if you actually pay attention. While these are called Mandela Effects.

Are solar eclipses genuinely created by an objective Moon obstructing the Sun, or they are simply assigned effects of their assumed but never actual, objective presence? Yet this is easy to find out. Just study eclipses yourself, very closely. Can you actually see the Moon during solar eclipses, or you may see only the Sun disappearing gradually away, by default, as caused by an assumed obstruction of sunlight by the body of the Moon? Is the Moon actually there during solar eclipses? Can you actually see it, and not only assume its presence as science does? Because you must study on your own these, while attempting to reach the main causes within their lines of causality.

No Sun and no Moon? But what exactly is going on in this world? And no Earth too? What goes on in this world is that this world might not be actually an ultimate reality as the current science claims now for over a century or so, but a created reality, either naturally created or artificially created, just as religion and spirituality claim for centuries and millennia. Along with millions of people who left this world and came back to tell their stories. Therefore, as it is the case in all created realities, not all details have to be present, but only their necessary effects and consequences, whenever these have to manifest. Yet whatever the case is, everything that is present in this reenactment, all details remain consistent with the entire world that they reenact.

And this is the case in all cognitive naturally created inner realities, because once these are part of a higher cognition, they have to remain relevant in all details and under all circumstances, for accurate results. Similarly, all your memories of your inner, subjective mind world where you reason must remain consistent with the outside world in all details, in order to allow an accurate reasoning. And this is why it is important to understand everything accurately in the first place, and not through consensual beliefs, stereotypes, ideologies, and entire jurisdictions.

Therefore, not everything that you may assume now about this world might actually be there, or at least not objectively, as many people and objects in this world, or as the people always present around. Because if specific objects, subjects, or even events, concepts, and circumstances do not have to be there through their objective presence or occurrence but only through their subjective, assigned, assumed, or expected presence and sequential implications, then they may be missing altogether from our world, leaving only their assumed, assigned, subjective presence to be there, as subjective image or information, as an illusion, or as simple decoration.

Take any object found around as an example. They are there and they exist objectively, since you can touch them. However, even these do not actually have to be made of molecules and elementary particles of any kind, since they do not have to be made of anything at all at a microscopic scale, just because you cannot see them at their microscopic scale. Because through their assigned, material presence, these objects are capable to give you their familiar shape and rigidity as they are at your own, macroscopic scale. All elementary particles composing them do not really have to be there, besides their effects at your macroscopic scale, giving you now the material, objective characteristics of your object. Search for all these molecules, atoms, and nuclei forming them, to find them nowhere, with only their effects affecting this world now, along with the objects that they compose.

And this is why you have to add ink whenever you want to see cells and subcellular components, because these might not be there. And therefore, you get to see your own ink now, shaped not by your molecules, but by their specific forms affecting the field around, not through their objective presence, but by default. Because it is the same with cells, molecules, atoms, and nuclei, they are not exactly there, but only the field is altered in their shapes and forms, and this is what you now observe. It is the same with many space objects, closer of further away, and it may be the same with Earth, since these do not exactly have to be there entirely, but only partially. As your ground and surface, or gravity, force of Coriolis, night and day, tides, and eclipses. While we may see throughout this book, how and why this can take place. Because if we already had accurate information coming from all space missions and explorations, then the Earth was entirely objectively real, and therefore entirely spherical, and we did not have to go through all these assumptions to model it now.

Who can help us throughout our research? Science can, and we already know all current scientific material ever since elementary school. True religion and spirituality always help, and I found them many times to be consistent with my results throughout all my models of this book series. While I never use their beliefs as initial conditions throughout my research, since I seek to use only accurate facts. Yet religion and spirituality are altered today by the same social actors altering science and education today. And this has always been the case throughout history, because all authorities of this world in one century or another managed to leave their customized influence in this unique world, and now every time we seek to find truth in this world, we have to be careful not to let their persistent revisions of this truth alter our results.

Yet we may get more help from those who exited this world entirely and then came back to tell their story. And we have millions of consistent testimonies to validate our results. And they do not state much about the actual shape and form of Earth, probably because they do not pay attention. Earth is spherical in the Etheric plane, and you may find anything about Earth there, all the information that you need. You may project in the Etheric in order to study Earth closely, to find it spherical, yet you find it spherical and flat, depending on circumstances. However, you may also find Santa Claus in the Etheric, along with Batman, Mario, and Spiderman there, and they may tell you that Earth is a sphere, or flat if you want, they can tell you everything that you desire. They will take you up in the air to see the Earth wholly if you want.

How much exactly is real in the Etheric plane? Everything is real and objective in the Etheric, as long as you are there, which is the case with all realities. While the Etheric is the summation of everything to have ever happened and existed on Earth, but only exactly as seen and understood by everybody on Earth without exception. In short, these are beliefs coming from everybody on Earth, and this is why you find Santa Claus and Batman there, along with what you did last summer exactly, and along with the entire flat Earth and spherical Earth.

Yet this is the case in all worlds and realities, and not only in the Etheric. And with our world being just another plane of existence itself, you may understand why Earth manifests sometimes as a sphere and sometimes as a plane, since it does so according to people's beliefs from one age or another. However, once the people of Earth may be capable to go up there in space to witness Earth as it truly is, everybody will witness the same thing, then and ever after, whatever they believe then that it is. Or more precisely, whatever the authorities manage to manifest by then through the multitude of people of Earth. And if they have a dynasty in the future, then it will be a flat Earth. Or it will be spherical, if nothing will change.

We have distinguished so far between beliefs, stereotypes, and accurate knowledge. We have also distinguished between genuine accurate knowledge and assigned knowledge by default, as it happens in the case of created realities, since created realities cannot be made in all details for lack of space, means, resolution, ability, and personnel.

We may also get help from flatearthers themselves

throughout our research. And we can consider flatearthers and planoterrestrialists right away, in the next chapter.

3 FLAT OR SPHERICAL, TRUTH OR DECEPTION

If you try to book a flight from a continent of the southern hemisphere to another continent in the southern hemisphere, you see how they fly you over this entire world in order to get you there, flying you through the northern hemisphere, unnecessarily. Now take an azimuthal equidistant projection map of this world, which is the basic flat Earth map, which is the map found on the cover of this book, take this flat Earth map now to see in amazement how you fly exactly according to it wherever you go in this world, probably because the Earth is flat and arranged in this exact manner. So here you have it, continents are displaced exactly according to the flat Earth model, and everybody flies exactly that way in order to arrive wherever they have to go. Right? No, not exactly, since boats seem to use a different map, the regular map you have on the wall.

Why hiding the true meaning of souls, incarnation, and planes of existence? In order to control and exploit endlessly the people of Earth, along with all souls. Why hiding that the Earth is another plane of existence? In order to make everything seem irrelevant, insignificant, unimportant, and very short lasting in this world, with your life spanning this existence on Earth, and nothing else. While it might be significantly different. But why having a spherical Earth and not a flat one? Why having to displace entire continents significantly on the map in order to make them fit a spherical model of Earth? Again, in order to isolate you from the rest of this world, since flat worlds are part of significantly larger planes and therefore larger worlds, while spherical worlds may be closed in themselves and therefore solitary in the entire universe. Do not underestimate the importance of this change in the model of Earth, since ten generations down the road, Earth may become flat entirely, or it may become spherical entirely, probably in order to match all spherical higher worlds, if those higher worlds are spherical and not flat.

Yet the Earth is not flat, but plane. It is the same, but it is different. All planes of existence are flat, since they cannot have any other shape as seen from the inside. While they do not even exist objectively as seen from the outside, since they do not have an outside. Yet there is more characterizing planes of existence besides their shape, and you should understand everything. Because if you do not understand realities entirely, you end up attempting to understand this world from one empiric effect to another, and this takes you nowhere. The Earth is plane or it is flat, but how exactly can it be so? How can the flat Earth generate its own force of gravitation? How can the Sun and the Moon rotate above Earth if it is flat? What exactly do these rotate around? Are these artificial too? Then how exactly can these fly above the flat Earth night and day? Is this entire world constructed mechanically? Questions are many to ask, and you can never answer them individually. Or you can, but you answer them empirically, in a simplistic, superficial, or erroneous manner.

Yet not all flatearthers are superficial, since many have scientific university degrees. They avoid topics that they cannot explain, as the force of gravity, Coriolis, or the Sun and Moon models, yet the rest of the model that they form for the flat Earth is relatively consistent, and it even challenges science with its own spherical model of Earth.

Are you a flatearther? It certainly matters when you read this book. Because if you are among its first readers, then you might be only searching for more information about the flat Earth. While if you read it further in the future, then you might already understand everything about Earth and this world. And then, calling yourself a flatearther or planoterrestrialist, is as calling yourself a Cybernauticalist just for surfing the Net casually. Yet how much of the flat Earth can you understand? Are you in fact more interested in why science fails in explaining everything regarding space and space objects? Then you might not be a flatearther, but you might be interested in this entire world. And even more, if this entire world spans as far as the lower orbit of Earth, making the Earth seem flat while doing so, then knowing everything about the flat Earth may be more important than everything else.

Am I a flatearther now? No, not at all, or at least not as everyone else, since I cannot side with anyone claiming that gravitational attraction is never related to Earth but to the difference in density, with denser objects falling down and with less denser objects raising in the air. And there are many other attempts that flatearthers make altering significantly a theory that could be accurate and could end up modeling and therefore explaining Earth, and through it, explaining this entire world. Yet many statements that flatearthers make are still accurate. And even more, if it was not for all flatearthers in this world, you never suspected that there is anything wrong with all scientific models that science offers, and you simply followed the current scientists in ignorance, with them lying to you the entire time. Since this is why they remove all space records that they ever 'take,' because people develop and can see through their lies. Just study the video recording from the Apollo 13 mission, if you can still find it. And they lost the original, by recording on top of it accidentally. So here you perform space missions of trillions of dollars each, but you have no proof.

How exactly have you become interested in the flat Earth?

Because it makes more sense, and now it is capable to explain most of the questions to have remained unexplained this entire time. As why you cannot find a fast, direct flight between Chile and Australia, and you always have to pass through Dubai, Los Angeles, or Chicago. Yet now you know it, because the Earth is flat, and all continents are displaced differently in the real world than what you learn in school. Just study your map of the flat Earth, or you may study the picture on the cover of this book, to see how, in order to go from Chile to Australia, you have to pass through America and Canada, through both South and North America, and through the entire northern hemisphere of the globe.

And this is one of the strongest case that flatearthers make. How does science explain this? Science is very consistent with society and with all its other domains. This is why you always feel as though it is always you against 'them,' because it is you against 'them.' 'They' are together against you. And this is how they get rich and powerful, while you become poor, helpless, and insignificant. Yet there is more in this world than wealth and power, since as stated previously, your significance in this world stands elsewhere, and it is only implicitly related to wealth and influence.

It is interesting to follow all anecdotes related to all cases that flatearthers make, since they are consistent. USGS, which is the American organization making all maps in this world, still has a budget of billions of dollars, which again, is more than enough to feed the poor from all underdeveloped nations, forever. What exactly do they do with the money today, when the entire Earth is already mapped? And why exactly do they choose, from all types of maps, exactly the azimuthal equidistant map, which is exactly the image of the flat Earth? Probably because this is the Earth. Or probably because they have to make the entire world to accept this model for Earth and the universe, for very specific reasons.

The return of the flat Earth, the return of the dark ages, and the return of the great dynasties of Earth. In short, all those tyrants and dictators of the West and of the East ruling the Earth from the hiding or in the open want dynasties reinstated in this world, since only throughout dynasties, people venerate them as deities. Because what they have today is not enough.

What exactly do pilots have to say about Earth and all its continents displaced differently than what they learn in school? Yet pilots get their salaries and do not have to say anything. Pilots never see UFOs either, since they lose their jobs if they ever do. And this is why this entire world is meant to perish, because the people themselves care more about themselves than they care about the entire world. Is this statement redundant? Do you also care more about yourself and your family than about the entire world? But do you care more about your family than you care about yourself? No. In fact, you sacrifice yourself for your family. Just wait until the next war, revolution, or major cataclysm, because you will sacrifice yourself for your nation and for the entire world, without a second thought.

Because everything that you do in life you do in order to fulfill your needs, while your needs come on various levels and classes. Currently, authorities force you to live your life on lower developmental levels and on lower class levels. As a reference, addictions are of the zero level, needs for servitude are of the first level, animal needs as physiological needs are of the second level, while intelligent human needs are of the third level. Your need to learn everything that you can find about the flat Earth is an intelligent human need, and it is of the third level. At the same time, needs that you fulfill on your behalf, are of the first class level. Needs that you fulfill for your family are of the second class level, while needs of higher class level address your community, nation, society, world, reality, and cluster of realities above. When your authorities become incapable to control you, as during wars, revolutions, or major calamities, then you return to fulfilling your natural needs in their entirety, and you do not have to fulfill only your lower level needs as you have to do today. It might sound strange now, but when authorities fall, your intelligent human needs become more important than your lower needs, and then you become eager to share everything with everybody else without discrimination, including food, shelter, transportation, and even highly important knowledge. Because all these make a better world, and making this world a better place is a very important natural human need. And when authorities fall, you manage to interconnect directly with everybody else, while you feel love for everybody else, without discrimination. I refer to this as horizontal interconnectivity, and it has everything to do with the flat Earth and with everything else, since everything is connected, everything forms the One.

This world is created and co-created, and through this specific co-creation, this world ended up to be exactly as it is today. And if you fail to understand the Creator and these specific higher beings along with this specific higher connectivity and this specific co-creation that they still work on, then you are doomed to reason in the exact manner that science and education have taught you to do throughout life. Because this world was not built or created as you build a large city, skyscraper, or cruise ship, but this world was created and is still co-created relatively freely and directly, by everybody, simultaneously, as they need or feel like, and this leads to all dualities in this world. Because you either have dualities in a world or you have harsh centralized directives, and it seems that you already chose exactly freedom, choice, independence, creative possibilities, and individual distinction and significance in this world, and not distinct, centralized, rigid blueprints and directives.

It seems that more and more people start doubting science and all its scientific explanations, just because information and technology become capable to allow people to find the truth in this world on their own. It is enough for you to own a pair of stronger binoculars, a telescope, or a more powerful camera, and you may succeed to see directly right under the curvature of Earth, in a beautiful day at the beach. Or you may take a mechanical gyroscope with you on a transatlantic flight, to see yourself following a flat course the entire time, as though you are flying above a flat Earth. If you are allowed with gyroscopes in commercial flights. Or you may use GPS technology throughout any intercontinental flight within the southern hemisphere, to see where you are the entire time, and to see if they take you through the northern hemisphere or not. If you get any GPS coverage there, or if you can ever find a direct intercontinental flight in South America, you find out that all restrictions throughout international commercial flights end up stopping you from learning where you are and how exactly continents are displaced on the surface of Earth.

It is the same with photography from very high altitude, since pictures can easily reveal not only the shape of Earth, which is what flatearthers always seek, but they can show you the genuine displacement of countries and continents on the surface of Earth, and this is why you can never find genuine pictures over the Net. Just search the genuine pictures that you can find, to see them showing a large amount of clouds, obscuring exactly what you are interested to see. And this issue fuels all debates involving flatearthers. While the laws of classical physics never allow most of these clouds to exist there, insulting your intelligence.

Because everything hidden and forbidden causes debates, altercations, and even stronger social movements. Because humans developed at their intelligent level have specific higher needs to make them fight hard for their freedom and for everybody's freedom in this world, for their rights and for everybody's rights in this world, and for highly important knowledge and for the availability of this highly important knowledge in this world. And as scientists, politicians, and all those composing the upper social level behave today, they do not actually seem to behave at intelligent human levels of development, but they strive to fulfill their lower level needs, their needs for servitude, along with their own animal needs, and this is exactly why this world is doomed. Yet all employed scientists and all politicians are in the Brotherhood, along with all astronauts and commercial pilots. And when you are in the Brotherhood, you do exactly what they tell you to do in the

Brotherhood. And now this is exactly your accurate knowledge. Since the Earth is exactly flat or spherical in this world, but only as the Brotherhood wants it to be. With all Brothers applauding and venerating this decision the entire time.

According to the flat Earth model, Antarctica seems to hold the very end of this world. However, all information about Antarctica is unavailable today. The military keeps Antarctica off limits from the general population, no one is allowed across it, and this makes flatearthers believe that there is a dome above Earth, with its limits in Antarctica. Admiral Byrd led expeditions in Antarctica several decades ago, claiming to have discovered an entire continent past the South Pole, as though he had managed to step into another world. And flatearthers still use his testimony to prove that the Earth is flat. Is this true or false? Can you fly across Antarctica to find the truth for yourself? Can you fly high enough above Earth to see its continents? Can you board a direct flight between two continents in the south hemisphere without taking you through the north hemisphere, and if you do so, then are you allowed to use your own GPS to track your flight? The answer is no to all these. And this is how you can never find the truth about Earth. Because they do not let you.

What can you do? Research some more, and as I always state, try to understand Earth and this world comprehensively, by understanding and connecting everything else, including life, created realities, the universe, the human reasoning, the human development, and much more, since all these are connected, and you can never have and understand one without having and understanding the rest.

What do flatearthers do? They also try to expand their research as far as they can, seeking the truth among all fakery offered in all space missions conducted by NASA. They depict all fakery found in videos and images everywhere. Wherever they find an image of Earth, they study it carefully, and if Earth happens to show an obvious square around it everywhere in all pictures, then that is exactly how it had been copied and pasted there. And then, throughout more elaborate research, flatearthers find out exactly where that image of Earth was copied from and pasted everywhere, or how scientists changed colors and clouds on it, or how they cloned clouds everywhere, or how they used the same layer of clouds in various photos, or how those same clouds remained motionless throughout entire video recordings of Earth from space. And everything is consistent, since science fakes its records. However, it means that space explorations and their records are fake, not necessarily because Earth is not spherical, which may also be the case, and not necessarily because Earth is flat, which may be the case. Sometimes, flatearthers found images of Earth with states as big as entire hemispheres. Other times, continents were smaller or bigger on Earth, depending on space missions.

What I noticed more recently is that science does not even record pictures and videos anymore throughout its newer and bolder space missions, but only spikes on a graph, as it did when it claimed to have landed a space module on a comet. Where are the pictures and videos of this outstanding human achievement? No pictures allowed, only spikes on graphs. Do comets even exist?

Yet this world goes on even with this entire fakery, and authorities keep controlling everybody including flatearthers, throughout similar fakery, in all social domains and not only in space exploration. While the social actors involved everywhere are always the same, and you may find them in finance, entertainment, commerce, science, education, medicine, government, politics, war, industry, and Internet, the same ones. While you also know exactly who these social actors are, controlling the entire world very tightly throughout all social domains, and you cannot say anything about it. Or if you do, you are charged with a very specific type of social and religious discrimination, enough to ruin your life, to land you in prison, and to render you significantly worthless and powerless in this world, with no one coming to your aid.

Because these specific social actors are strong, and when

they can summon the entire world against you, you will certainly feel it. Because the main issue about this world is not that it is flat or spherical, it is not that flatearthers are ridiculed and therefore discriminated, but the main issue is that the entire world is a prison, acting as an extermination camp, and you can do nothing about it.

4 SUCCESSFUL REASONING AND CONSENSUAL CONSTRAINTS

How exactly should this research be done, if the current explanations are empiric and superficial? Even though it was the observed curvature of Earth to be inconsistent with the current scientific model of Earth, or even though relevant records are missing from all space missions, this does not mean that we have to start our research of the flat Earth and spherical Earth exactly with them, since these are simple effects. Their main causes are further up on their lines of causality, and we have to seek exactly these main causes to study. And when we do so, it is possible to jump from one social domain to another, and therefore we have to be prepared through our understanding to model these. Because as stated previously, you have to understand not only this world, but you have to understand all realities, all creators, all intelligences, all life, and all connectivity. This entire knowledge is absent from science, and therefore you have to find it yourself. This book series happens to hold dozens of models helping us throughout our research of the flat and spherical Earth theories, yet what we find is more than a simple theory, since we have to find everything relevant correlated to the flat or spherical Earth. Throughout this chapter, we construct this specific line of causality in order to keep us oriented throughout the model. More precisely, we start with the knowledge offered by flatearthers, and we try to move causally up the line toward the main causes.

We start with the fact that, throughout time, and throughout this world, Earth was flat. It is claimed that two ages ago, in the Age of Aries, they used a spherical model of Earth in particular cultures, yet even then, this world was flat everywhere else. China, kept a flat model of Earth even during the first part of the Industrial Age. Therefore, if you are surprised of the significant interest that the flat Earth presents today, most of it relates to religion and spirituality. And this detail is so significant, that as stated previously, it might be powerful enough to return this world to the type of religion and veneration that have taken place during the dark ages, and long before.

Is this hard to believe? Because these social actors showing up throughout this book and throughout all books of this series are capable of everything, even to turn this world back to the dark ages, with the Earth flat, with them inhabiting Earth on their own, and with you and your entire genetic line long departed. But then, who exactly is going to do the veneration? Who exactly is going to have to go throughout all the misery of the future dark ages? They will, while those that they will venerate are not even human.

But is the Earth flat from the beginning? Can these actors accept a spherical Earth from the beginning, and therefore go against the model of this world offered by their own religion? Because it is easy for you to find and study their religion, in order to find the truth. And what you will always find is that this truth is what they have been always keeping away from you, unless you are one of them. And it does not matter if you are a scientist, a politician, a businessperson, or a member of a lodge, you have to be part of them genetically in order to find the truth, because they have all the necessary means to know and manipulate the truth. Otherwise, you cannot prove that

Earth is flat, spherical, or in any other shape.

We also notice that science lies in everything related to the shape of Earth. Not only this, but the biggest and the most guarded scientific lies address space, the nature of our world, Earth as a whole, and your own meaning in this world. This has been the case during all ages, dark or not, and this is the case today. And this is exactly why science is dogmatic today, behaving just as any of the tens of thousands of cults and religions of this world, not because it is incapable and it covers its ignorance with lies, not because it pockets the money as NASA does with its very large budget, but because it has a very important meaning in society, to hide all relevant knowledge that could ever free humanity. It keeps you in this manner ignorant, underdeveloped, and indoctrinated, as it does with its entire army of scientists, teachers, and professors. These remain indoctrinated, and in turn, these indoctrinate this world, including you and your loved ones.

And now, moving forward throughout the line of causality, we find that all scientific theories are fake, erroneous, misleading, and therefore harmful for you and for this world, including the big bang theory, the theory of evolution, quantum mechanics, and the theory of relativity. And then, moving outside physics, we find psychology to be harmfully erroneous, fake and misleading, meant to harm you and your loved ones, diagnosing you with mental sickness if you happen to be more developed, as it happens with all children labeled with autism. Medicine is faked, drugging you and getting you high while it claims that it cures you but it does not. It only kills you slowly with poisonous medication and with harmful radiation. Finance is meant to bankrupt you, education is meant to indoctrinate you, fake religion and fake spirituality are meant to enslave you and your soul, economy is meant to waste goods and therefore make you work all life long, industry is meant to trash this world, democracy is meant to discriminate this world, capitalism is meant to impoverish and exploit this world, and justice is meant to accuse, charge, and punish you regardless if you are guilty or not, while all these are meant to empower them, and all these are meant to determine you to enslave and eradicate this world yourself, on their behalf.

Because they are not allowed to work, they are not allowed to fight in wars, and therefore nothing that is wrong in this world is their fault, but it is your fault, you are making this world as it is. And this is why higher beings will never come to your rescue whenever your time comes and you and your genetic line have to depart, leaving only them around. Until they have to depart, since nothing is about them in this world from the beginning, regardless of what they are promised, but there are beings of all kind waiting in line to have Earth and this entire world for themselves.

All flatearthers come as far as identifying these social actors as main causes, yet they fail to predict future religious implications of the flat Earth. They fail to see the flat Earth theory as a significant part of an entire religious model of this world soon to come back and rule everybody or at least those still around. And to go further on this line of causality, powerful forces are very determined to reenact all religious records concerning the end of this world, and these span not only science with its official flat Earth theory, but everything else, including final divisions of nations and religions, decisive wars and sacred places where they have to fight again, with plagues and atrocities succeeding exactly as described by religious records, down to the last trumpets calling all these important religious events. And these already happen, one by one, they take place, you are made to reenact every one of them, and you do so diligently. Such an extraordinary theatrical performance, in such a comprehensive sacrifice.

But why doing so? Do not blame anyone. Just look around, and if there is anything that you do empowering them to go ahead with this majestic plan, then you are the one to do so, you are the one empowering them, you are the one using their money, working for them, indoctrinating yourself, sending your loved ones to be indoctrinated, and harming those around. And you do all these through beliefs and lower level thinking.

Therefore, you have to understand everything about accurate facts and beliefs, how they act on your cognitive system, how your cognitive system works through them, how you think, how you reason, who you are, how you connect with others, and how you behave in this world. And as stated above, you have to know everything about this world, including your meaning and place in this world, in order not to act against Life and against this world in everything that you do.

This world is a plane of existence as the true religion and spirituality claim. Why a plane of existence and not a sphere of existence? I distinguish between vertical and horizontal connectivity in this world throughout all books of this series. Planes of existence tend to have all life connected horizontally, one with another. Spheres of existence have life connected directly both vertically and horizontally. As a reference, families are connected horizontally, and this is exactly how mothers know when their babies are hungry, because they are connected cognitively subconsciously or even consciously, and they may remain connected cognitively throughout life. This type of horizontal connectivity was supposed to manifest throughout this world and not only within families, yet it is just as ridiculed as the flat Earth theory.

As a reference, this type of cognitive horizontal connectivity associates with telepathy, while all psychic abilities are ridiculed and banned by society. Only movies are accepted, since movies are nice. However, you may be capable to continue your intelligent human development to include all higher abilities, and not only telepathy. And this is the case just because all higher beings have all their higher abilities, and even souls have these higher abilities, since they are higher beings. Animals have higher abilities, and they use them subconsciously. Humans have them, and they use them randomly and subconsciously for themselves and for their family, many times against themselves and their loved ones, while society is capable to control them in particular ways in order to use human higher powers continuously, as part of their indoctrination. I refer to this social scheme as the milking and the churning, and it always takes place, everywhere.

We have arrived at the human development, and at how humans are made to remain underdeveloped throughout life, through drugs, irrelevant lifestyles, austerity, lack of true, pertinent, accurate knowledge, and through the continuous poisoning, done by harmful medication and food additives. Because everything happening in this world has the same meaning and leads to this same result: genocide, the end of this world, enslavement, veneration, the return of the old dark times, and the same higher entities to return to Earth and rule some more, as deities. Can we model all these?

5 EXTRAORDINARY LIFE IN AN EXTRAORDINARY REALITY

Humans are extraordinary, and it is even more extraordinary to have billions of humans in this world for decades and centuries now, with none even suspecting that this world is flat or plane this entire time. This tells how minutely society is capable to control and indoctrinate this world. Did you actually assume that centuries and millennia ago, people were strongly indoctrinated to believe that this world was flat? No, since it might be the other way around, so let us see.

I refer to all planes, realms, and spheres of existence as realities, along with all worlds and universes, since these are different realities. All realities form the wider world, or the One, or the Deity, or Life, since all these are defined to be everything that exists, depending on the perspective that you adopt. As a reference, living perspectives lead you to Life, as everything alive that exists, or to Intelligence, as everything intelligent that exists, or to the One, as everything and everybody that exists, since they are connected. Or connected to the Creator, to the Deity, or to the Divine, when you perceive everything from religious perspectives.

Our world is everything that exists, objectively, spanning as far as its spacetime continuum is capable to hold and define everything and everyone that exists objectively. No other reality exists objectively outside our world, or if it did, it was part of our world. Similarly, you can never escape our world in any manner. It is the same with all realities, since they are objective in nature, but only while you are within them. And therefore, for them, from their perspective, nothing exists objectively outside them, not even our world.

Realities are everywhere, yet you do not have to find the end of one reality in order to exit it and therefore to find another reality to enter. This is impossible by definition, while it is impossible practically. What happens is that realities are capable to form other realities within themselves, and I refer to these as inner or lower realities. Soon, you will be able to identify them. Similarly, our world is formed, held, and maintained in a specific matrix within our higher reality. It might be the same with our higher reality, since it can be held, formed, and maintained within its own higher reality, upwards, up to the ultimate upper reality, up to the One.

And this is possible, since existence has three relative natures as seen from our world. Existence is subjective, objective, and highjective, depending on where you are and on what reality you are in.

We are going to create an inner reality right away. Let us now form a matrix, and through it, let us form a specific continuum capable to hold and define everything in our newly created inner reality, since these kind of inner realities are everywhere. You may do so by creating a digital code in any computer, since this is how all digital matrices are formed, capable to hold the computer operating system, which is the digital continuum, capable to hold and define any object, subject, and event of any computer world, as it happens throughout the videogame world of "Mario", or throughout the videogame world of "GTA5," depending what you like to play. Or you may create inner realities by simply daydreaming, since the neurons of your brain and even the regular cells of your body are capable to form the necessary intelligent encoding through electric or electrostatic fluctuations anywhere within the brain and within the entire organism, intelligent fluctuations capable to form the intelligent matrix that is capable to hold all continuums of all inner worlds of your cognitive system. And these are countless, existing subjectively within your mind, forming your mind, while also holding the multitude of your inner intelligences helping you reason now throughout your study of the flat Earth, and every time throughout life. Out of this multitude of cognitive inner worlds of your mind, we will choose one specific inner world to study here, since it is the most significant one. And this is your intelligent inner replica of this world, or your inner world, spanning the cortex. But before we do so, we have to continue defining realities in all the necessary details.

Let us see now what exists and what does not exist here, and in what exact existential form they do. In fact, everything exists in the One in any existential form, since if it does not exist in any form, then it is not part of the One. It is almost similar with our world, since everything to exist objectively is part of our world, as everything that you may see everywhere, with the exception of everything that you imagine and everything that you may see on the display of your laptop or TV, because these exist subjectively, and therefore they exist in this manner only within inner realities formed by our world.

The hardware itself of your laptop is objectively real, and it is therefore part of our world. It is similar with all wires and all electricity going through all wires, since electricity, light, all radiation, and all electric, magnetic, and electromagnetic field are part of the field and therefore part of the spacetime continuum, part of this world. However, the encoded, digital fluctuations formed by the successive opening and closing of all switches of your motherboard only forms a digital matrix now, holding and maintaining within it a new, inner reality. And there is where Mario lives, along with the rest of the artificial intelligences, in this computer inner reality. You cannot go there in your physical body, Mario cannot come here in its subjective body, and this is the case just because you cannot transfer objects and subjects from one reality to another, but only personalized copies of these, including personalized copies of this information. Digital inner realities happen to be very accurate in transferring information. However, mind realities are different, just because mind realities are very dense in details. As a reference, you have to start drawing if you ever want to transfer information from your daydreams to our world, or you have to start telling the story, because there is no other way, since you cannot transfer genuine information from one reality to another.

We have four types of worlds that we have covered so far: computer worlds, mind worlds, this world, and our higher reality holding our world in a matrix. Science claims that our world is the ultimate world, as there are still people in this world choosing to believe so. In fact, you can never know if our higher reality exists, just because if it exists, it has to exist highjectively, since it is our higher reality. And since it is impossible to transfer information directly from one reality to another, then you can never transfer higher knowledge directly to our world. Therefore, everything coming from our higher reality is formed of only interpreted copies of this information. And this is exactly why you always have to rely on the teachings of others, as on the teachings of religious characters, of spiritual characters, or of gurus, just because they cannot transfer directly higher knowledge to you and to this world, but only interpreted information, whatever they are capable to understand and explain themselves. It is almost the same with the millions of people to have departed this world and then returned to tell their story, only that they went there in person, through their higher selves, and witnessed in this manner all higher circumstances by themselves.

Who exactly is who here? Mario is subjective in nature, along with all your daydream characters, and all movie and book characters, which are fictional. You are certainly objectively real here in this world, along with all objects and subjects around. While your souls are highjective in nature, and not subjective as religions and spirituality try to understand them. You are subjective for them, while they exist highjectively, along with our entire higher reality, along with all higher beings from all higher realities, and along with the Deity, or Life, or the Divine, or the Supreme Creator.

As stated above, you cannot leave any reality to go to another. You may do so, but only through another self of yours that happens to be already there, in that other reality, as it is the case with Mario. You may interact with its game world through it, and therefore Mario is now your lower self or digital avatar in that game world, or your lower body. It is the same when you daydream, since you do so through your inner self if you want to interact with your daydream in the person of your inner self, or you may simply dream in general without you present there. However, you are not always using the most powerful mind in the known universe in order to daydream all day long, but you use it to think and reason, depending on circumstances. You undergo your social reasoning through mental models, more specifically, through the enactment of real events and circumstances, through genuine social simulations resembling common soap opera, and I refer to these as mental models.

I am going to define a reality now and I am going to use mathematics to do so. Because mathematics applies to other realities alike, while physics applies only to our world, defining here all the natural laws of this world. I refer here only to genuine physics and mathematics, which are mostly the physics and mathematics of the past centuries, and not the fakery produced by science today. Why the difference? Because throughout the current world order that started before the world wars, everything produced by society is irrelevant, fake, and misleading.

And it is significant to know who these social actors are, since they are not the current consensual, hierarchic Brotherhood, but the invisible kingdom. Which is an entire nation coming from the Caucasian region in Asia, now found all over this world and counting in hundreds of millions, ruling this world systematically from the top of all its scientific domains, while served closely by the current consensual, hierarchic Brotherhood.

Here is the definition of any reality, along with its characteristics: a reality is a distinct set of objects and events spanning a commutative topological ring or a commutative topological vector space. A reality might also be a set of more than its continuum, objects, and events, or less, depending on circumstances.

In abstract algebra, the words 'commutative,' 'topology,' 'vector space,' and 'rings,' are sufficient to assign to a set or to an entire world the following properties:

The set or reality is continuous, boundless, and it never reaches its limits.

You may never exit or enter any reality in any manner.

You may not go from any point A to any point B of any reality and pass through the outside of that specific reality, or through a gap in reality. Therefore, there are no gaps in any reality and no exit and entering points.

Each reality has its own sets of natural laws, operators, and operations defining all criteria of existence of that specific reality.

There always exist null elements and inverse elements within a reality as defined by all internal natural laws and operations specific to that reality.

And these are enough to define realities. Note that these are not beliefs but facts, relating to the natural laws of this world. You may reason through them in order to apply them to anything you need. If mathematics is too tedious to follow throughout reasoning, we may define realities in simpler manners, as:

All realities are objectively real as observed from within, while nothing exists objectively outside them, not even other realities.

You may not assume that other realities do not exist according with their own separate laws of existence.

Realities do not intersect themselves objectively. Realities do not share information, components, and matrices of continuums in an objective manner.

You may transfer only interpreted copies of objects, subjects, and information between realities, and nothing else.

Note that any model of the flat or spherical Earth has to follow these laws.

You always interact with other realities throughout life, with countless of them many times simultaneously, and you do so while using computer technology, while reasoning, mental modeling, daydreaming, reading books, watching movies, or listening to stories, while you may also interact with other realities while dreaming, astral projecting, and even while projecting continuously in any lower world, as souls do even now as you read this book, since you might be one.

Therefore, as you notice, it is always through you that you manage to exit and enter realities. You do not do so directly, but through already existing bodies in all these realities, subjective and highjective bodies that you may refer to as avatars, other selves, souls, spirits, and lower and higher bodies.

Note that these are you, just because you live your life not individually, but in an entire lifeline of existence, having on it all your selves, living life many times one through another. All your selves live life in a multitude of worlds and reality, with one for each world or reality. Because as I always state throughout my books, you are greater, more complex, and more diverse than entire worlds and realities, because you are an entire lifeline of existence containing all these worlds and realities. And probably you notice by now how insignificant science and society make you be and feel throughout life, while it is the other way around.

But who are you? You are mind, body, and soul. You are intelligence, spirit, and soul, because you are alive, you are intelligent, and you are the physical body defining you objectively in this world. Because in each reality where you exist, project, and live your life, you have to have an objective body or self. You have your physical body defining you objectively as who you are here in this world. You have your inner self defining you subjectively as who you are within your own inner replica of this world, where you reason and daydream. You have your soul defining you highjectively as who you are in your higher world, defining you similarly in all higher worlds where you exist highjectively above this one, if you do so. And you have all your inner intelligences defining you in all your inner cognitive worlds of your mind, inner intelligences that you use continuously throughout your thinking and reasoning.

And now you may understand life, since life is the physical body along with all intelligences that it holds within its inner worlds, and therefore life is always lived at the confluence of realities, along an entire lifeline of existence. In your case, your life is lived at least at the confluence of our three main realities: the higher reality with your soul, this world with your physical body, and the inner reality of your mind with your inner selves and all your inner intelligences helping you daydream, mental model, think, feel, learn, develop, and reason. Therefore, you are your mind, body, and soul, you are the three at once in one lifeline, at least, because you are more complex, spanning countless of existential lines and clusters of realities.

Who are you exactly? You are the three at once, mind, body, and soul, yet these three selves form your cognitive system, since your mind is divided into three major parts: your subconscious, conscious, and highconscious intelligences. Out of these three, you consider yourself the conscious intelligence. And this is the case because technically, you do not have your soul, but your soul has you. It is as Mario claiming that it has you, which is the other way around, you have Mario. Yet you may still claim that you have a soul, just as you may claim that you have a city and a nation, while you do not own them, but your city and nation have you.

Therefore, you may tend to identify yourself with your conscious intelligence, out of all intelligences of your cognitive system, just because you cannot access other intelligences directly. You cannot interconnect with your eating intelligence, and start one day digesting all proteins that you had for breakfast, consciously, one molecule at a time, since this is what your eating intelligence does, all day long. Your entire subconscious is formed of primal subconscious intelligences, as your eating, reproductive, recovery, or social subconscious primal intelligences, among others, and these intelligences send you your needs. You also fulfill your own conscious needs, as you fulfill your own developmental need right now by reading this book. You are even rewarded for being able to fulfill this need, and you can feel the happiness rewarding you right now, since it is induced through serotonin.

Your highconscious intelligence sends you your higher needs, and you have to fulfill these. Which exactly are these higher needs? Just pay attention throughout the day and throughout life, to identify them. Just do not substitute them with lower level needs, as entertainment or addictions, because they feel different. Yet if you have never fulfilled higher needs before, you may assume that they feel the same. And do not use beliefs and ideologies in order to find your higher needs and higher purpose in life, since these are meant to control you, your soul, your loved ones, and the souls of your loved ones, through you and through your determination to use dogma and beliefs in order to find or give a higher meaning to your life. Because this is exactly what people do, and look what they do to this world. And by considering this entire genocide taking place for centuries and millennia in this world, you might be tempted to assume that beliefs and stereotypes are harmful.

We will model your learning, reasoning, and mental modeling shortly, since these are highly important to understand. You can already understand your place in this world, through your mind, body, and soul. Yet what you want to understand is your meaning in this world, or at least one of your meanings in this world, a most important one, as a creator of your own inner reality, which is your own inner replica of this world, where all your memories and understandings of this world are.

You notice how you cannot go in the orbit of Earth in

order to see this world wholly, in order to see its shape along with the true displacement of all nations and continents on the surface of Earth. Yet you can manage to understand all these once you understand how you learn and reason. Can it be possible? Can our model be capable to model the human reasoning, and then can it be capable to link it directly with this world, with Earth, and with the shape of Earth? Let us see, yet you might have already figured out everything by now.

As stated previously, everything that you do throughout life you do in order to fulfill your needs, while throughout your entire cognitive system, you are the conscious intelligence. In fact, all intelligences are conscious, as slight as they may be, just because all intelligences have to interact with their immediate outer environment in a successful, conscious manner. Intelligences may exist independently throughout realities, yet they tend to gather and form cognitive systems. Just as cells gather to form organisms, people gather to form cities, nations, and societies. I refer to these as classes of life, and they may come on many levels, since all living beings tend to live life together, in classes of life that are of various levels.

There is a difference between living beings and the intelligences of these living beings, since all living beings have a physical body and an intelligence, which is in fact a system of intelligences, a cognitive system. However, there are intelligences throughout realities capable to transcend physical bodies, mostly in order to evade death and destruction. These are mostly higher level intelligences, yet you have a multitude of these intelligences within your own cognitive system, and they are capable to transcend to the new bodies of all newer generations, at the moment of conception. I refer to all these as primal intelligences, since they never die, and they have been alive since organic life and way before. If you ever feel highly determined to engage in all sexual activities, this is the case just because all your primal intelligences desire to remain alive after you die, through all new generations. And it is a form of genocide to stop you from reproducing, through the multitude of irrelevant laws, rules, and regulations stopping you from finding a partner, stopping you from engaging in the act of reproduction, and then stopping you from giving birth. And do not worry about overpopulation, because it is only a myth. Because the upper society manages to reproduce abundantly, while the lower part of society has significantly fewer children. And this is how their genetic lines gradually fade away, while the upper classes prosper significantly, and so they take over the Earth. And this is genocide, taking place alongside wars, austerity, sickness, misery, fabricated illnesses, and fabricated terminal illnesses, as everything is meant to kill everybody today.

And among all these primal subconscious intelligences, you are a primal intelligence, and you call yourself conscious, because you can identify only yourself as being conscious, from within the entire cognitive system. And just as all primal intelligences are highly specialized, in eating and digesting, or in reproduction, recovery, security, or excretion, you as a conscious intelligence are responsible with the interaction of the entire organism with the outside world, while you fulfill your needs on behalf of all primal intelligences, of all cells, of the entire cognitive system, and of the entire organism. And on behalf of your entire family, nation, society, and civilization.

Yet these are only your natural needs, since all dogma, stereotypes, control, servitude, and ideologies give you more needs, consensual needs, which you have to fulfill throughout life. These are your artificial or consensual needs, your forcefully imposed needs, and they certainly take time and resources away from your natural needs. And if by not fulfilling your own natural needs, it causes you to become sick, old, tired, disabled, or addicted, by not being able to fulfill your natural needs of higher classes, as your family, community, national, and social needs, it causes this entire world to end up disabled and enslaved. And this defines entirely this world as you find it when you are born, and then as you leave it behind when you die, since everything is neglected and therefore controlled, dysfunctional, and dualistic in this world.

Yet there are other intelligences within your cognitive

system, since you and all your primal intelligences have to reason through these other intelligences, and this is how you open lower realities to reason within, through these lower intelligences. And then these lower intelligences also have to reason themselves and so they open newer lower realities to help them reason, down to basic algorithmic procedures, as if-then and repeat-until. And together, you break down cognitively all information coming from the outside world, into elemental cognitive bits of information that you are capable to isolate, understand, and relate to everything that you already know, and then store it or memorize it within your databanks or inner worlds, for a later use. And you store them there not exactly as databanks, but you create in your mind an entire living replica of the outside world, with all these memories. And this is how you store them, in the shape and form of the outside world, as you have found them in the outside world.

And how was our world formed? In the shape and form of the higher reality. And everything is alive here, just as above, and it is cognitive in nature, from the perspective of this world above.

You notice how, throughout this same line of reasoning, all inner intelligences manage to create their own inner realities, inner replicas of their immediate cognitive environment, where they have to reason, and where they are the main, genuine creator, just as you as a conscious intelligence are the creator of your entire inner replica of this world, where you reason right now, as you read this book. Because all intelligences from all realities are alive, conscious, and certainly intelligent, and they think and reason through and within their own naturally created realities, while they are the creators of these realities themselves.

Note that this specific line of thought or reasoning does not start with you, but through the specific circumstances of the outside world where you are made to live your life, you are actually continuing higher lines of thought or reasoning, and therefore this is another one of your important meanings that you have in life and in the wider world. Which is of a cognitive nature, meant for you to take part in this extraordinary line of thought or reasoning spanning entire existential lines of realities, with you, with your inner replica of this world, and with this entire world as a significant cognitive existential chain of reasoning within the entire higher reasoning procedure. And this is how you have comprehensive, higher lines of reasoning matching higher lines of causality throughout comprehensive lifelines of existence, which are the living beings themselves, as they live their lives throughout their multitude of worlds and realities, many times simultaneously, with this world only one of them, and with you only one of these living beings.

I refer to this as comprehensive, analytical reasoning, and it is what you should always do in order to have an entire cognitive system based on genuine accurate facts, and not on beliefs. Because what beliefs do, they enter your cognitive system wholly through pure memorization, you fail breaking them into basic accurate knowledge, you end up thinking through them and not reasoning through them since you cannot do so, and this is how you become incapable to understand this world, you live your life from one problem to another, you have to depend on others that have to depend on others throughout entire social hierarchies of power and control, and so you live your life, and this is this world that you make. And this is the case just because you do not understand the difference between analytical reasoning and thinking through beliefs, just because education and psychology failed to explain them to you, wasting your time and effort with indoctrination, entertainment, and irrelevant information. This is the case with everybody, and in this manner, everybody leaves open the multitude of consensual existential niches of this world that allow other beings and other entities to come here and rule this world, through the ignorance and servitude of the people of this world.

And as you may notice, any mistake that you make throughout life and throughout your reasoning affects all your higher selves and all your higher realities, since they live and reason through you and through all your lower selves or intelligences, which also depend on you and on your understanding of this world, on your reasoning, and on the structure, consistency, and accuracy of your inner replica of this world. If you feed them beliefs and not accurate facts, this causes their inner worlds to resemble work camps, factory farms, and military units. This is where they think, and now this is the kind of consensual thoughts, needs, ideas, impressions, and feelings that they give you. This is how you live your life now, and this is exactly the kind of thoughts, needs, ideas, and impressions that you send above, to your soul and to your higher world. Because the Consensual Matrix extends itself everywhere it is allowed to operate, in your inner and outside worlds, and in your higher world, always through you and through the rest of your selves, higher and lower. Because you either fulfill natural needs for Life, or you end up fulfilling artificial, consensual needs for the Consensual Matrix. While it is your choice, and it has everything to do with your learning, development, and reasoning.

Why exactly do you have to reason and learn? Because you have to learn how to behave throughout life. And everything that you do in life you do in order to fulfill your needs, from the zero level addiction needs, to your first level servitude needs, to your second level animal needs, and to your third level intelligent human needs. What happens is that all intelligences of your cognitive system have to tend to their own tasks, zillions of them, throughout the body and cognitive system. And these include all intelligences of your cells and subcellular components. Whenever these cannot fulfill their tasks, they send their needs to you. There is not enough water in your digestive system, it is too cold or too hot, you have to visit the bathroom, your neighbor is too good looking today, and you have to fulfill all these needs as much, as fast, and as well as you can. Or you can simply watch some more cable, drink some more whatever you drink, and so you postpone your needs.

And then, there are needs that you cannot fulfill, which is the case with your entire human and higher needs, since society ignores or even ridicules human and higher needs, just look around. You cannot fulfill these, and consequently, your intelligences punish you with pain, misery, boredom, and depression for your failure, and these certainly push you to drugs and entertainment, if you are not careful. Yet you still manage to find useful books and podcasts today, so you can still find a way to develop and therefore escape drugs, servitude, and divertissement.

And this is how you behave throughout life, how you fulfill your needs, and most importantly, this is how you have to reason the entire time in order to know how to fulfill your needs. You receive your need, you figure out how to fulfill it, and if it is successful in your mind, you apply it in practice. Or if it does not work in your mind, then you have to reason again. And so you have to think some more, and you try it again, and again, until you have it done. Or you ask others to help, you search the Net for solutions, and so you find your way to solve the problems and fulfill your needs.

Through mental models, you are capable to fulfill your need in your mind first, see if it works there since you do so faster and with no effort at all, this is called planning, and then if everything is viable, you apply it in the real world. And if it works, you call it a successful solution, or a successful idea. Then you memorize it, you improve it, you use it repeatedly, you share it with others by bragging and chatting continuously, and everybody does the same. And you did so in your mind first, through a mental model, just as you would have done in the real world, and it worked. Mental models are part of your reasoning, among many other cognitive processes.

You may use mental models before you start repairing your car or house. You may use mental modeling before solving social problems in the real world, or you may use highly advanced mental models in order to understand very tedious and very delicate concepts referring to the entire world, just as you are working on this specific mental model of the entire flat or spherical Earth as you read this book.

Mental models are part of analytical reasoning, and they

help you understand these elementary bits of information coming from the outside world, which I refer to as accurate knowledge, accurate facts, or accurate truth. It is easy with classical physics and mathematics, since you learn most of it in schools and universities, while it is harder with everything new and tedious to understand, as the model of the human reasoning itself, or the model of the flat Earth.

What you want now to understand is how you perform your mental models, who exactly reasons throughout your mental models, and how. Yet you reason yourself, consciously, through your inner self, in your inner world, most of the time, while you are helped by all the other intelligences throughout your mental models. They reason themselves within their own inner worlds, in about the same manner, depending on their specialization and therefore depending on the topic of your problem that you have to solve, since analytical reasoning is a cooperative cognitive activity. As stated previously, you are the conscious intelligence within your cognitive system, and you reside in your left prefrontal cortex if you are right-handed. You live there within a very small group of neurons, and in yourself, you are actually a cellular intelligence. It takes you an entire group of neurons in order to live, reason, and fulfill your tasks, because you need a variety of connections within the entire brain, organism, and cognitive system. While you also need space for all your cognitive inner abilities or inner intelligences to perform your tasks and fulfill those needs. And from there, from the prefrontal cortex, you spread out to reach your senses of perception, you look outside live and you hear what goes on, you walk around, you stretch, you feel everything, and so you assume that you live your life at the macroscopic level of your entire organism, while you are only the conscious intelligence, small enough to fit in one neuron.

Notice how there are two of you, the conscious intelligence, and the physical body, two selves, two you, but only one, since you the inner self, interact with the outside world through your outer self, which is your physical body. And by accessing your senses of perception and your muscular system, you are capable to project from the inner reality of your mind to live your life outside, in the real world, as your entire physical body. And these are your two selves, your inner self, and your physical body.

And this is exactly how life is lived at its basic forms, with one physical body and one intelligence held by this physical body in an inner world. Yet life is never lived in this most basic form, but life spans a multitude of chained realities, with a multitude of selves, bodies, and cognitive systems in an entire lifeline of existence, with all its selves sharing a same purpose, all fulfilling the same needs throughout life. And it is similar with you, since you have a multitude of lower selves and higher selves, forming your own lifeline of existence, throughout the multitude of your inner, outer, and higher worlds.

This world is only one world among the rest, it depends on the worlds above just as its inner worlds depend on it, and therefore the higher worlds form, maintain, and change it exactly as they need. While this world changes, forms, and maintains its own inner worlds exactly as it needs, or exactly as you need, or exactly as everybody needs. Therefore, it is not exactly you changing this world one belief at a time through direct manifestation as stated previously, but the higher selves perform these changes in this world through their own cognitive abilities, just as easily as you can make similar changes yourself in any of your inner worlds. And if you want to change the shape of Earth in your mind from flat to spherical, you simply do so. Yet everybody has to do so, all souls, simultaneously, and only if the Creator of this world allows it.

The difference between the people of Earth and their higher selves is that souls tend to be more developed, while they also have higher abilities, as telepathy. Since telepathy helps them have large, comprehensive common worlds that they form, maintain, share, develop, and change together, as this world.

Would the people of Earth like to create similar common cognitive worlds themselves? They cannot, because some of these higher selves had decided one day, long ago, not to allow humans to form similarly shared common inner realities. This is how humans do not become as them, and so they scrambled their languages, or they tempered with their cognitive abilities. There is no telepathy for humans now, and therefore there are no commonly shared inner worlds. And this is why now you have to get on the Internet in order to fulfill direct interconnective needs, because you cannot do so directly in a common mind spanning society.

And now, you may understand the meaning of this world and of all living beings and intelligences interacting with this world including humans, souls, and even higher beings and intelligences, since everything and everybody is part of all realities involved in this entire higher living and cognitive procedure. This is why souls come here to live life through you and through everybody else, not because they actually enjoy to come here as it is portrayed by everybody today, but because all souls and all similar higher intelligences above them are part of the same living and cognitive activities and of the same chains of life, intelligence, and existence, taking place within relatively similar realities upper and lower. With higher realities being wider, more accurate, and more consistent, and with lower realities being smaller, truncated, more specialized, less accurate, and less consistent.

And now, we may answer the question of how Earth can be spherical and flat at the same time, and the answer is that out of all details, events, and sceneries of all higher realities consisting the main image of the replica consisting our world, this world is only one chunk, one specific specialized truncated scenario of the higher reality, with our specialization taking place at the surface of Earth entirely, at least for now, and therefore it does not have to involve Earth as a whole, as seen from above, along with the immensity of space beyond, and all space objects. We may continue this reasoning to state that our specialization here in our world is mostly of a social nature, here at the surface of Earth, and it probably has an entertaining nature, for now.

Is Earth flat or spherical? Our specialization does not include this detail, simply because humans do not need this specific detail in their lives, for now. Even more, humans do not need to know anything beyond Earth, since they do not have to interconnect with anyone beyond Earth, for now.

But why, if this kind of knowledge is so meaningful for the human reasoning, for the human development, and for the human condition? There is no answer here coming from above, yet you already have your answer, since the same beliefs, ideologies, and social stereotypes that are in the above world are in this world. Because the above world is the one that created and that maintains now this world, and they create this world exactly as they believe, exactly in the image of their beliefs, even consensually. And now, this is what their own beliefs state, that Earth is spherical, and this is what they create. Therefore, the same beliefs are here as they are up there, and as they are within all inner mind worlds of all humans from Earth. And this is why there is nothing beyond Earth today, nothing above this world, and therefore there is nothing to consider, just work, work, work all day long, drive, watch TV, interact with others, take drugs, be happy, and nothing else. But one day, when the higher selves decide to understand more, and therefore decide to develop more, there will be space around Earth here in this world, along with stars, planets, and other civilizations. While the Earth will also have a shape, and it will be probably spherical, since the laws of classical physics point that way. Or who knows, if the Consensual Matrix decides otherwise, then all higher selves decay, fall back into dark ages, again, along with this world and the multitude of lower realities from everybody's mind. There will be servitude and veneration then, directly for the Elite and for all entities controlling the Elite. All stars will rotate around Earth once a day along with the sun and the moon, and the Earth will be flat all along. Or who knows, there will be no Earth at all, not anymore, if up there in the higher world, the Elite decides to hunt down all our souls and burn them at the stake for their telepathy and the rest of their higher abilities, since this happens often. Because you are always a better servant in the Consensual Matrix as a soul, as a human being, and as anything else, without your higher abilities and higher senses of perception.

And you notice how you participate in all these scenarios through your own beliefs, reasoning, and decisions, causing even the decay and death of your higher self in his higher world. Because when the humans of Earth use beliefs in everything that they do, then they remain at the mercy of their authorities, as these do everything they please, and so it happens in all higher and lower worlds. But when humans reason, behave, and interconnect at their intelligent human level, this is exactly what their higher selves will do up there in their higher world, they reason, learn, and develop. The higher world becomes larger in this manner, to include stars, planets, and other civilizations, if these really exist, since they learn everything that is in their own higher world. And this is the case just because your behavior, decisions, and achievements are their thoughts and reasoning up there, and through them, this is what they know and how they behave. And this is exactly how you have a choice in all these, including in the shape of Earth, so just make it a good one.

How exactly do you reason and form your mental models? You have to learn first, since you need to know everything about the problem that you have. Whenever you encounter a problem, you have to study and know everything about it first, and then with all the data in your mind, you have enough to perform your mental model. And you simply go through the entire procedure of solving your problem in your mind first, before you have to do so all over again in the real world, if you happen to find the successful idea. If not, then you have to think some more, surf the Internet, or ask around, and they give you the successful idea already validated, along with how to use it, and it has to work.

What is the difference? In the first case, when you struggle yourself to solve your problem, you happen to develop cognitively significantly while reasoning and mental modeling yourself, and so you advance in development. While in the second case, you have to depend on others to help you from now on, because you remain underdeveloped and you are certainly incapable to solve the rest of your problems from now on, since problems tend to get harder to solve throughout life, and not easier. And then, when you become entirely incapable to solve your problems, you lose your car, house, and job, you lose your friends and your family, you get sick or you have to retire, you have to go to the hospital or to court, to jail, or even to morgue. It happens to everybody eventually, but only when they stop developing, and consequently, only when they stop solving their own problems. While science remains incapable to provide you with the accurate knowledge of how to solve most of your problems. It is the same with education, psychology, and medicine, because you fail some more or you have to work hard and learn throughout life in order to become the scientist, the teacher, and the doctor in one, just to help yourself, finding your own way out. Because society harms you and exploits you, but rarely helps you. And if you take drugs, well, good luck to you.

How do you make your mental models and how exactly do you reason throughout them? You have to learn everything relevant first, as stated above, and you do so just by studying the problem yourself, by learning and understanding everything in the outside world.

How exactly do you learn? You do so throughout life, gradually but continuously, since you learn not only at school and in the family, but you learn through all your experiences, through all your interactions with those around, from everything that you watch on TV and read in books, or from the Internet. If you are not careful, you end up learning through beliefs and stereotypes, if you do not break everything apart into cognitive elements when you learn it, down to elementary pieces of information called accurate facts. This is called understanding. And then, after you break down the knowledge, you have to elaborate it, which is, you have to link it with everything that you already know. Otherwise, if it does not fit, if it is not compatible with your previous knowledge, then it is not an accurate fact or accurate knowledge, but it is erroneous, irrelevant, misleading, or even doctrine, dogma, and therefore harmful information. Or it may be the other way around, what you learn is accurate knowledge, but you cannot accept it and therefore you cannot link it with your previous knowledge and understanding, because your entire cognitive system is filled with beliefs, and now your newly encountered accurate knowledge is incompatible with your own beliefs, and you discard it fast. Study society now, to see how it provides beliefs, and it encourages you to think through them. The common language uses the words believing and debating as forms of human thinking, and even psychology does the same and considers lower forms of thinking and behaving as acceptable human thinking and behaving.

And this is how you know to distinguish accurate knowledge from stereotypes and beliefs. Or this is the case when your memories are already based on accurate knowledge, because if your memories are based on beliefs, then you cannot distinguish between accurate knowledge and beliefs. What you can do, just start studying classical physics in every manner, either through examples or through pure theory and applications, and you get there. And then try to understand your car, computer, TV, smartphone, and the entire network, and you make your way toward understanding the entire world. This is certainly better than taking drugs and getting entertained. While it is significantly better than spending your life in servitude.

Your entire bank of memories is not actually as a library of data, since only computers have these, but your entire memory resembles the real world closely, since your memories are in fact your inner replica of this world. Just look around now to see your room outside, and to find your room in that exact shape and arrangement inside in your mind, within your inner replica of this world. These are your memories of your room. Because this is what you do throughout life, you are actually constructing or creating this inner replica of this world through learning, understanding, and apprehending of this world. You move in this manner all knowledge and information from the outside world into your mind, you create your entire inner replica of this world yourself through learning, experiencing, and understanding, cognitive brick on top of cognitive brick.

While as you notice, you are the creator of this entire world, the conscious intelligence, along with your entire lifeline of existence, with all your selves higher and lower aligned, since many times, they participate in your life as they please. And therefore, you are the creator of your entire inner world. While everything is created in your own image, or in your own talent, characteristics, details, constraints, limitations, expectations, desires, and understanding of this world. And this is the case with all selves and intelligences, since they have to construct their own inner replica of this world to reason in order to fulfill their needs, and this is what they do.

And the more successful you are at learning, the more your inner replica of this world resembles the outside world. And if your ideologies teach you that the Earth is flat, or spherical, or in any other shape and form, this is exactly what you believe and what you learn, and therefore the Earth from your own inner replica of this world becomes exactly spherical or flat, whatever they say. And this is exactly what your inner replica of this world will have from then on, a flat or spherical Earth, since this is what you have learned in school. And this is exactly what your inner intelligences will have to know from then on about Earth, that it is flat or spherical, exactly as you know it from the outside world, either as an accurate fact, or as a belief. Since as you notice, this entire detail or cognitive brick is transferred throughout your lifeline of existence in this exact manner, from above to below. Yet it can go the other way around, since the worlds below reason for the worlds above, and now this is what they think.

Who are these inner intelligences? Throughout life, while you learn and experience everything, you learn everything about everything, you store it in your inner world exactly as it is found in the outside world, and this includes not only all objects along with everything that you know about everything, but everybody just as well, the inner replicas of all the people from the outside world. Because these are in your inner world exactly as you know them and exactly as you believe them to be in the outside world.

These are not simple dolls or mannequins there, but your own inner intelligences animate your inner replica of all the people that you know in the outside world in all details, exactly as your own conscious intelligence animates your inner self in your inner replica of this world, you. And since all intelligences are alive, all your inner replicas of people are alive, in a subjective manner. Or this is the case from your own perspective, because from the inner perspective of your inner world, all your inner intelligences and all concepts and conceptions there are genuinely alive, since existence defines them objectively at their own level, as long as they are there. And while all characters are the physical bodies there, the intelligences impersonating them are their cognitive systems. And it is the same with you the conscious intelligence and your inner self.

And it is the same with you, since you have your own inner self at the interior of your inner replica of this world, consisting of everything that you know and learn about yourself, the one from the outside world. You are also within your inner replica of this world as an inner self, at its center, since throughout your life, you have understood and therefore replicated yourself continuously in your inner replica of this world. This is actually your inner self. You as a conscious intelligence have linked yourself directly, rigidly through axons and dendrites with this inner replica of yourself from your memories, your inner self. And this is exactly how you live your life now, assuming knowingly or unknowingly that you are your inner self, and not the conscious intelligence. Or this is the case only when you choose to live your life within your inner replica of this world as your inner self, as you do throughout your reasoning. Because when you choose to live your life in the outside world, you do so through your muscles and senses of perception, and so you become your outer self, your physical body, the one that you see in the mirror. And this is exactly how you live your life, projecting repeatedly in the inner world when you want to think, reason, mental model, daydream, feel, imagine, contemplate, plan. And this is how you mental model. And more importantly, your inner concepts and conceptions of the outer people are the ones mental modelling for you, through their normal inner lives there, in your inner replica of this world.

Because when you have to fulfill your needs, you get out and project in the outside world, in your physical body, and you are ready to go and roam the outside world where you may interact with anyone in every manner, where you may learn and experience in any way you want, while respecting all rules and regulations. While you have managed very well so far, as a cellular intelligence.

How do you learn? You form and consolidate your inner replica of this world continuously, replica formed of all images, understandings, concepts, and ideas of the outside world, all present in your inner replica of this world. You may interact in this inner replica of this world in any manner you wish, as your inner self or as anything or anyone you wish, and this is exactly how you reason, mental model, feel, imagine, and plan. As seen, you replicate in your mind the real people from the outside world, and once these end up in your mind, they end up there as the inner selves of all those around. They end up as real intelligences in this manner, real and free to roam your inner replica of this world as they please. This is exactly how they live their lives there, and this is exactly how you conduct all your mental models. You simply offer the new, necessary circumstances, you form the truncated inner scenery in this manner, and all the inner intelligences of your inner replica of this world are simply living their lives there, in any circumstance you need them to interact and behave throughout your reasoning, exactly as you have to fulfill your needs in the outside world yourself, or as you daydream, or as you write your books, or as you reason alongside this book. And then, whatever they achieve at the end of your mental model and entire reasoning, constitutes the final, successful idea.

And this helps you plan and predict in the real world, in the real society, among the real people. Since if you are very skilled to learn and understand everything and everyone around, they behave as naturally and as accurately as possible in your inner world, and this is exactly how you reason and mental model throughout life. And the better you are at learning, the more accurate your inner replica of this world is, the better you are at understanding even more knowledge, continuously throughout your experiences, the more successful you are throughout your reasoning, and then the more successful you are in the outside world. Because you are already capable to understand everything through accurate facts, exactly as it is in the real world. But most importantly, you cannot achieve it in a higher level analytical manner, at the intelligent human level, if you do not know all these, which is, if you do not know exactly how you reason in the first place.

Are these intelligences within your inner replica of this world actually real? They are just as real and just as alive as you are, the conscious intelligence, only at a lower existential level, the subjective level, yet still real, alive, and certainly intelligent, from their own objective perspective. Because everything is objectively real in every reality, as long as you are there, and therefore everything is objectively real within your inner replica of this world, for all your intelligences, since your inner replica of this world is a genuine reality, even if it is an inner reality. Besides, this world is an inner reality in comparison with our higher reality where the souls live. While you as an inner self there within your inner replica of this world are a genuine, real, intelligent living being. You simply live your life casually there, and this is exactly how you think and reason throughout life. Life is a stage in your inner world, you are actors there, fulfilling continuously a cognitive purpose, and this is exactly how you reason.

And for your entire inner world, you are the great creator, the great architect. You are also the conscious intelligence in your own cellular world, while within your inner world, you are the inner self, your own avatar there. It is the same in the outside world, where your avatar is the physical body. While your soul lives high above you in our higher world, as you are its avatar here in our world, you, the entire organism, along with the entire cognitive system.

Yet as stated before, it is only a matter of limited perspective to interpret multidimensional life in this manner, divided into selves and realities. Because from a higher perspective, your life is lived simultaneously through all your selves and realities, and even in one same place, with all your selves, higher and lower, superimposed in one. And probably with all similar realities superimposed in one. And this is the case only because we are capable to interpret Existence only through its three separate natures: subjective, objective, and highjective, while all these natures are superimposed in one existence, as seen from the perspective of the ultimate reality, from the perspective of Life, Intelligence, and the One. You are mind, body, and soul, as one.

How accurate exactly is your inner world? It depends, since people are better or worse at learning and understanding this world. And what is always worse, you tend to be convinced that your inner replica of this world is entirely accurate, and things cannot be in any other way. As a reference, just take a pencil now and start drawing whatever you see around, and that is exactly how perfect and accurate your inner replica of this world is. And some people are more or less talented at drawing and understanding this world than others, while others spend their entire time addicted, entertained, or lost in the never-ending soap opera present all around. Because people live their lives without ever understanding how they think and why. They cannot understand all their selves but they assume to be their physical body, they do not even know that they have a replica of this world, it is filled with beliefs, stereotypes, and discrimination most of the time, and this is exactly why we have this world that we do, literally, and this creates now the duality of our flat-spherical world.

How accurate is your replica of this world? Just learn and understand everything through accurate facts, keep developing throughout life, avoid drugs, entertainment, and divertissement, fulfill your intelligent human needs as you do now, and it can become very accurate.

Not that you will ever get a prize or anything from society for your successful reasoning, intelligent human behavior, and continuous fulfillment in life and in this world. They might even remove you from this world for all these, since this is this world today, under the control of the Consensual Matrix.

While you also notice that your mind and therefore replica of this world would have never been compromised, if the Consensual Matrix would have not infiltrated it with its beliefs, laws, stereotypes, rules, strong convictions, jurisdictions, and entire ideologies. And it is the same in the outside world, since this world would still be fulfilling and therefore meaningful for Life, Intelligence, and the Supreme Being, if it was not infiltrated by the Consensual Matrix in a similar manner. And it is the same everywhere throughout the wider world.

And now, you know how you learn and how you reason, while you can also understand naturally created realities, since all intelligences are conscious, specialized, and they learn and reason in the very same manner as you do, by creating their own specialized replicas of their immediate world, and by reasoning and mental modeling within them. And this is exactly how all intelligences open newer, lower realities throughout their lives, fill them with knowledge and inner intelligences, cognitive brick by cognitive brick, learning everything from their immediate environment, everything necessary to perform all tasks throughout your body, within all organs, within all cells, within all cellular components, zillions upon zillion of inner intelligences and inner worlds, inner realities where even smaller inner intelligences roam and interact freely, all living their lives casually, not even realizing that their world is very little and it has a cognitive nature. Because from their inner perspective, within their inner worlds, everything is objective, happening just as naturally as possible.

And this is how you receive your needs, feelings, and ideas from all these intelligences, you use them to reason even better and to fulfill your needs, and through your needs, you end up fulfilling their needs, since everything that you do in life you do on behalf of your cells and all intelligences of your cognitive system, zillions in number. And as I always state throughout my books, you are greater, more diverse, and more complex than entire worlds and realities.

And now you have the necessary knowledge to understand this world, along with its entire nature and meaning in the higher world. However, our world is different than our higher realities, just because humans lack higher abilities. Because humans are consciously intelligent while interacting with those around, while interacting with the rest of intelligences within the cognitive system while reasoning and learning, and while fulfilling inner needs. Yet humans were supposed to be consciously intelligent while interacting with their highconscious mind, allowing them the use of higher cognitive abilities. And this is why humans are different than souls, because humans cannot connect directly, telepathically with those around, with everybody around, in order to form larger, common inner mind worlds, where they may interact freely within their own common inner realities. Computer technology promises to offer this direct, technological cognitive connectivity, while humans were supposed to have it naturally. Because those replicas from your inner world of all those around were supposed to be linked directly with their real counterparts, while they are not.

And therefore, you are not a telepath now, you are not connected directly with those around, you are isolated, alone, vulnerable, and therefore controllable. When those higher beings came down to Earth ages ago in order to scramble people's languages, this is exactly what they did, they ended up disconnecting everybody, rendering them isolated cognitively, on their own, thinking alone, feeling alone, and living alone.

Because souls from our higher reality are different, they are capable to connect with each other, as they may even form extraordinary inner realities, extraordinary inner worlds where they may interact, reason, mental model, and daydream together in any manner they please, and it must be interesting. Because you do not have to tell stories anymore in the outside world, you do not have to watch stories in the outside world, but you may simply be together in one mind world watching the same line of memory, the same sketch of imagination, contributing to it as much as you can with your part, with your own imagination alongside others, alongside your loved ones. And it must be beautiful, daydream after daydream, one beautiful scenery after another, one precious moment after another, life after life. And now you may understand this world, since it is one of these co-created realities, made even right now as you read this book, by billions upon billions of souls. And if you just look around, you may see exactly what they did and still do in this world. This is what you do, what we all do, just look around to see it for yourself.

There are many types of naturally constructed realities, and you may work on them your entire life throughout your learning to detail and perfect them in order to help you reason, or you may descend within them to live your life there entirely if you please, as you may do throughout dreams and astral projection. Or if you happen to co-create your inner reality alongside others, then you may live life after life there, within your common inner world. Yet co-created realities are more tedious and more delicate to form and maintain, since everything changes within co-created realities according to everybody present, you are not always in control since it is not exactly your own inner reality, you work on one portion of that co-created inner reality and not on the entire thing, and this is exactly how you end up with dualities and discordances, just because everybody understands life, this world, and their meaning in life and in this world in different manners. You may have rigid, predefined co-created realities where everything is meticulously perfect and consistent everywhere, yet you risk to lack diversity, ingenuity, complexity, and uniqueness, and these are the most precious qualities in this world that every intelligence seeks and treasures. And this is why you always prefer free interactive co-creative realities, for their extraordinary diversity, ingenuity, and spontaneity.

And now you understand all dualities from our world, coming from the fact that people understand this world in various manners. In the first place, people live their lives at the surface of Earth, and therefore they assume it flat, even unconsciously. And then, only when they fly around, they expect it to be spherical, which it cannot be anymore. It is the same with people and with all souls co-creating this world, with some knowing more or less about the actual laws of physics and mathematics, and this is exactly what they now do. Our world is significantly denser than the multitude of your dreams, denser even than the multitude of astral planes where you may project. And now, for the first time, you have noticed the first dualities and discrepancies that this world presents, and it must feel striking.

Is this world flat or spherical? Well, it is both, depending on references, and depending on people, just because people are powerful enough to create this world in any manner they want, but mostly in any manner they can, as it is always the case with your own inner replica of this world. And if you were wondering how accurate your own inner replica of this world is, now it is significantly more important to know how accurate this world is. Because as you see, at its highest, comprehensive scale, it does not actually add up to offer you the same consistent world that you experience, enjoy, and take for granted down here at its surface, throughout your casual existence.

6 THOSE IN CONTROL OF THE HUMAN KNOWLEDGE

Why have you never learned all these? They hide this knowledge, they censor it. Who? Everybody, all major bookstores and websites, since they own the Internet. Yet in fact, this is not exactly an internet, but another façade, since all computers in this world are not connected freely as the word 'internet' claims, but computers connect only to big servers and from there to big websites. Nobody sees anything wrong with this picture, this is exactly how all knowledge and connectivity in this world is still censored and controlled many times heavily, and this is how this world still remains indoctrinated. And this is what the question if this emerging flat Earth theory is a natural movement, or it is strategically implemented right now in these highly important times, for various reasons. And we have to find these various reasons now, starting with their main causes, or you will never understand anything in this world, if you follow independent indices. Because you have to understand the entire agenda at play, who or what stands behind this agenda, and what these social actors implementing it today are promised and will finally get. Because lies manifest on all social levels, and not only within the Masses. While you have to be able to distinguish them from among all truth, wherever you are positioned in society. Because the entire humanity is in it, and through it, the entire humanity is vulnerable, rich and poor alike. And with the rich killing the poor in a major genocide, humanity becomes more and more vulnerable to anyone or anything up there.

Is this really a failure from the part of our souls, to have created such an imperfect, unjust, and discriminatory common world? This world is correspondent to the higher world, making the higher world similar to this world. Yet even as imperfect and unjust as this world seems to be, it is highly enjoyable, highly fulfilling, and it seems to be highly successful. Souls have worked hard to plan all these in slight details, and this is what we now have. Well, then why don't souls come down here in person to experience everything firsthand, so they can enjoy this genuinely harmful world exactly as they want it? Yet they always do, and they love it. How exactly do you like your videogames? Fast, rough, thrilling, violent, harmful, and realistic.

What exactly do souls want in a created reality? This. They want exactly this. And it is exactly the same with you and your movies and videogames. Do you watch beautiful movies with colorful butterflies flying all around under the beautiful blue sky filled with little puffy-white clouds floating nicely all around? Yes, certainly, you watch them for about a minute or so, and then you want to get back to your war videogames, since they are highly intense and they keep you engaged for weeks in a row. It is the same with this world, since it keeps everyone engaged in a natural manner for life. Since this is why it is nicknamed Terra, probably because it gives you the teror and frights that you always expect.

We are going to make a short model of society in this chapter, with everybody involved in the fakery that takes place in the entire world, and this includes all fakery involving Earth and its true shape. We also note that if we understand society accurately, we have the chance to understand it from a variety of perspectives simultaneously, from within, from above, and from within all social classes and all groups of social actors involved, because the specific topic of this book allows it.

Who exactly is 'them?' The rich, certainly, since the rich control the entire world, including science and education providing all knowledge in this world. And why can't they just come out one day to tell bluntly: 'this world is flat and we have lied this entire time?' Because there will be major riots in the entire world forever from then on? Well, no, nothing will ever change, but only a few lines in a textbook, and this is all.

Who are the rich of this world? You might be one, if you have managed to land a good job in society, making over one hundred thousand dollars or euros a year. Because this is how sociology defines all social classes, by the amount of money that people make a year. You are in the bottom social class if you make under twenty thousand dollars a year or so, you are in the middle class if you make up to one hundred thousand dollars or so, and above that limit, everybody is rich and very rich. Therefore, if you were looking for all responsible social actors, now you have just found them, since these are the scientists, famous actors and writers, all dentists and stockbrokers, and all owners of successful businesses.

Well, but this is just another lie, because you can never get any of the jobs and business opportunities mentioned above, if you are not in a lodge, in a mob, dictatorship, business cartel, radical political party, or in the invisible kingdom. While any information starting with this level of awareness in this world is censored, and everybody knows it. What exactly is hidden here and what is left in the open? Whatever it is, this makes the difference throughout all riots and revolutions taking place in all nations. Because when people have enough injustice, austerity, fakery, and indoctrination, they start protesting against no one at all, since all social actors in control of this world remain hidden, so no one can address them. And this is called ruling this world in the open, as it happened during the dark ages, or as it has to happen with false deities, because in order for this world to venerate these, this world must know them first, they must be in the open. While ruling from the hiding gives you the advantage of remaining safe and therefore ruling longer, but you never get to rule as a deity, but only as a hidden dictator, hidden tyrant, or hidden megalomaniac, with no one knowing of you.

For example, the current presidents of the West are not actually leaders or rulers of their nations, since others above own these, from the hiding. These are the actual rulers, but you never know them, since they are in the hiding. And it is not the same enjoyment for them. Because all deities are venerated abundantly, and so should they. And it is not possible in the hiding.

And this is where the flat Earth stands, because through its striking presence, it is capable to bring back the harsh, authoritarian religions of the dark ages capable to maintain under control all Masses indefinitely, while assuring that the Masses can venerate their ruling deities indefinitely.

What is the difference? There are not only the Masses venerating the false deities, but their souls do so alongside them, and this makes the greatest difference in the wider world. Therefore, while modeling the true society, we have to consider all these premises.

As you notice, there are entire social classes hidden in society. Yet to be more precise, only the upper social class, the Elite remains hidden from the rest of society. While the middle social class, the Brotherhood, may remain in the open, while its entire activity and meaning in this world remain hidden from the bottom social class, the Masses. And this is how, whatever sociology models erroneously into its three social classes: the poor, the middle, and the rich, well, this actually applies only to the Masses and the Lower Brotherhood, since the rest of society remains hidden. As a reference, if you happen to live your life casually exactly as you learn in school and as you see on TV, then you are part of the Masses. If you happen to be born in the Brotherhood, then you might not read this book, since the Brotherhood offers its own knowledge to its members. If you have just joined your local criminal cartel, your local religious organization, or your local lodge, then you are in the Lower Brotherhood. And get ready, because you will make everything happen, from the bottom of the entire world. As a reference, presidents and ex-presidents are in the Middle Brotherhood. The Rockefellers are at the top of the Brotherhood, along with all ruling families of all nations, everyone owning the central banks of their nations, and implicitly those nations. While the Rothschilds might have already made it in the Elite.

The Elite is made of the few top families of this world, yet through intermarriage, they form a genuine larger human family now, while multiplying extensively. And do not expect the same rules found within the Masses to apply to the entire world, since the upper social layers reproduce freely and exponentially. While the Lower Brotherhood is allowed to have more children. The Middle Brotherhood have as many children as they can, legitimate and illegitimate. Large numbers strengthen all families of the Upper Brotherhood, since life is lived through families from that social level up, and there are no reproductive rules involved, certainly not the rigid rule of having to be married with one partner for life, or with one partner at a time, or having to be married at all, or having to respect specific ages and other conditions, because these are only rigid means for population control. The Upper Brotherhood and the Elite use technology at its best in order to reproduce, and this is how it is no telling how fast they reproduce and how soon they are capable to replace this world on their own.

What offers stability to this world? Is it wealth? There is an entire ideology in this world, the capitalism, offering the basis for this entire civilization based on capital of all kind, yet money and wealth covers only the two bottom social classes, leaving the Elite to live their lives entirely above money. And with the Brotherhood controlling the Masses entirely on behalf of the Elite, then why exactly would the Brotherhood accept to serve an entire upper social class, and not to eliminate it and replace it at once, and take over this world? This happens constantly throughout history, yet the Upper Brotherhood and the Elite are genetically related today, with the 'pure' genes up among the Elite, and with the 'inferior' genes among the Upper Brotherhood. And this fuels the entire eugenics movement throughout this world, resulting in this continuous, extraordinary genocide. Yet the upper part of society has humans that do not consider themselves humans anymore but true deities, and they expect and demand to be served and venerated as deities. And the Upper Brotherhood serves and venerates them, as deities. The Upper Brotherhood does, since the Lower Brotherhood remains ignorant of anything found above the Brotherhood. And for these people, the Earth is flat, since for them, the dark ages have never ended.

While the Masses are maintained in control exactly as you know it. Why don't the Masses revolt? They do not know against whom to revolt. And this is how every time they riot, they break windows, their windows, and they turn cars upside down to set them on fire, their own cars. The Masses also engage the police and even the military, while these are also part of the Masses. The Lower Brotherhood is ordered to participate, in order to control all riots to take place exactly in the manner demanded. And this is how you get to watch on TV how the police sprayed protesters with water and pepper sauce during riots and got in trouble, or how the children of specific dignitaries participated in protests and turned one police car upside down and sat it on fire. And this is how the Masses tackle any injustice in this world, through ignorance.

Yet the Masses are highly capable, so can they ever gather the means to overtake the Brotherhood, to buy their businesses and networks and just replace them, increase wages, give everybody a job, and therefore make this world a better, equal place? No, not at all. Because capitalism engages this world in a genuine race. And this entire race is asymmetric from the start, just because the Masses compete individually, while the Brotherhood is organized into highly consistent hierarchic cartels, which interconnect to span this world, helping in this manner one another. Therefore the name the

Brotherhood. Because the Brotherhood does not include only the masons, but it includes all business cartels of this world, along with all religious cartels, political parties, and criminal cartels. All these are the branches of the Brotherhood, legal and illegal, they always cooperate throughout a continuous channeling of wealth from the Masses up toward the Families of the Upper Brotherhood controlling all nations of this world and all central banks. And therefore, the Masses stand no chance to overtake the Brotherhood, while they never intend to do so in the first place.

There is still the hypothetical scenario with the Masses uniting entirely in a larger family to take over this world in this manner, as in a revolution scenario, making everything equal, but the Masses are almost extinct, since most of them are already genetically eradicated, with the rest transferring fast to the Lower Brotherhood. The Brotherhood is the new Masses in the dark ages, as they are always the perfect conscious servants. Compared to the Masses who have no clue.

And yet, the Masses get their way more and more today, while the Brotherhood hurries to exterminate them, one genetic line after another. We have to figure out what is going on here, and mostly, why don't the upper social classes use brute force one day to draw a decisive line in this world between who is here to serve and venerate indefinitely, and who must die right away.

Because as you notice, there is more taking place, just because this world is not as isolated and as insignificant in the wider world as you may expect, and it already has a well-defined meaning in Life and in the wider world, which cannot be fulfilled otherwise, but with everything remaining exactly as it is today. Just because this world is made in the correspondent image of our higher reality, where the souls are. And these live life in servitude, these are the Brothers and the Masses, as these control everything, as they are also controlled in the higher worlds.

But first, let us now see how the Masses understand society. They certainly see the Brotherhood ruling this world through very tight hierarchies, while they also see something else, the invisible kingdom. As they cannot even define it as a separate genetic line, but only as a major religion. And it is not sufficient to understand their exterminating intentions in this world, as they control this world minutely. And this is exactly how they control this world, or at least the majority of nations of this world, and this is how they trigger, control, and profit from all wars in this world, while they are not even allowed to work and fight in this world. They are never to be blamed for anything happening in this world, yet here they are, at the very top of this world, subjugating this world, committing all indoctrination, all control, and all genocide in this world. And if you ever study them, they are the ones claiming that they are the most harassed, exploited, and exterminated in this world.

And these are the major social actors at play in everything happening in this world today. These people keep you underdeveloped and astray today, and with everything happening in this world, from wars to poverty and financial crisis, these are the people that you have to study in order to understand the major agenda behind everything. And they are certainly behind the flat Earth and spherical Earth, and you have to know everything, or you risk going extinct.

But how exactly is the invisible kingdom capable to take down and control the mighty Brotherhood, and even the mighty Elite? As stated, the invisible kingdom was an actual real kingdom in the Caucasian region until only centuries ago, when it migrated in mass to Europe and throughout the Commonwealth, currently ruling the West entirely.

It is only a matter of social competition, as society allows and enhances this specific type of social competition today, with wealth, influence, and material interests highly prioritized. And just as the Masses can never compete socially with all social organizations of the Brotherhood, now the Brotherhood cannot compete with the invisible kingdom in a very similar manner, just because the invisible kingdom counts in hundreds of millions, all following strictly a mandatory agenda which is also their tight ideology and their religion, and the

Brotherhood stands no chance. And with the upper social class of the invisible kingdom now in the Elite, the Brotherhood really stands no chance. While this is a true kingdom, as kingdoms normally are, with all the people of the invisible kingdom in complete servitude and truly devoted to their royal family above.

But let us comprehend this world very well now, as it is divided into the Masses, Brotherhood, and Elite, with the invisible kingdom in the Brotherhood controlling it tightly, and in the Elite, doing the same. Because if you are in the Masses, you know everything, but you cannot react. While in the Brotherhood, you cannot react either. If you are the CEO of any corporation, you are faithful to the Brotherhood, since this is why you are a Brother, and this is why the Brotherhood made you a CEO. While if you are a mayor, senator, or president, you are always faithful to the Brotherhood, since you are a Brother, but you are not faithful to the people voting for you. And this is the case everywhere, since even the people that you hire to work in your business are faithful to the Brotherhood, since they are always Brothers, but not to you and your business. And you know it, since you cannot be from the Masses and own a business, mostly hiring Brothers, with you in the Masses. It is impossible. Or it is possible to hire Brothers if you are in the Masses, in very rare occasions, and expect the Brothers that you employ to take over your business.

And this is why the Masses remain subdued to the Brotherhood. While it is similar in the higher worlds and throughout the wider world, since the Consensual Matrix is everywhere in the wider world, at the underdeveloped level. With the same privileged Brothers and Elite everywhere, since this is why you have an entire Consensual Matrix in the wider world, to make the Brotherhood and the Elite possible.

But now, if you are in the Brotherhood and in the invisible kingdom, you are always faithful to the invisible kingdom, and the Brothers cannot do anything about it. Since the invisible kingdom is your actual genetic family, it is your actual normal family brotherhood. And so you are always faithful to your invisible kingdom, while exploiting the Brotherhood on behalf of the entire invisible kingdom. Since this is how you rule entire nations in the West when you are in the Brotherhood and in the invisible kingdom both, with the Brotherhood serving you flawlessly. Since the invisible kingdom owns the money in the West, along with all profitable business, along with all public sectors including all governments, and along with the entire police and military.

While in the East is different, since the East has removed the invisible kingdom, having their own dictatorships in the open, while seeking to expand in the entire world. Through war if necessary. With the invisible kingdom worthless the entire time. Because even in the West, the invisible kingdom hides behind the European native genetic lines and behind the Brotherhood, incapable the entire time. Only that the invisible kingdom owns medicine, and through it, it eradicates the other genetic lines, seeking supremacy in this world. And with the tyrants and dictators of the East seeking the same in the entire world but with them in power, now this is the current world war.

Where do you stand in rapport with the invisible kingdom? If you have ever marked yourself as a Caucasian on any official form, then you are now part of the invisible kingdom, at its bottom, and you are the one working and fighting on its behalf. Even more, the invisible kingdom owns you just as you own your shoes, just because you have marked yourself as a Caucasian whenever you have applied for your certificates, documents, permits, ID cards, and licenses, since you have branded yourself to them, officially. Yet the invisible kingdom considers that it owns the entire world anyway, with you included.

Yet just because you are alive today, along with your entire genetic line, it means that you are kept alive and that you are still in good terms with the invisible kingdom. Because the unfortunate nations and genetic lines to have ever bothered the invisible kingdom are extinct today. And this is exactly why you find the unnecessary dictators and mass executions for the last century. While they use a centralized computer system to keep evidence of all the people in this world, where they live and of what genetic line they are, and now this is how this world ends up with terrible illnesses and dies, going extinct entire genetic lines at a time. And this is how the genocide goes on.

In short, the invisible kingdom eradicates the other genetic lines at least in the West, and now they remain exposed and vulnerable, since they were hiding behind the other genetic lines the entire time. And it is only a matter of time before the Brotherhood stops serving them. And with the tyrants and dictators of the East taking advantage, it might be too late.

Why would anyone in this world ever want to be part of this? It is not only an order or a constraint, but it is a need determining you to give these terrible orders on one side, and to implement these orders in minute details on another, to cooperate at once with all ruling parties and social classes, or to make you accept your faith naturally. Because everything seems science fiction now when you read about wars, ideologies, outrageous discrimination, and even continuous genocide, yet it certainly seems so because you happen to live your life in a favorable mode of life, and not in one of the multitude of crisis modes of life. Because in any of the crisis modes of life, you happen to receive extraordinary needs as the need to survive under any circumstance, and you become a genuine animal then, constrained to find any way out in order to survive, and nothing else.

Or this is the case if you happen to live your life on lower developmental levels, because at the intelligent human level, you help those around. And if you help each other in this manner, you manage to get out of any dreadful circumstance. But now, with you genetically eradicated, nothing applies to you anymore.

From an upper perspective, once you live your life drugged, under servitude, and through animal instincts, you will always become a dictator yourself if you have the chance, a tyrant, a megalomaniac, and even a deity demanding strongly to be venerated continuously by your people, unconditionally, with their mind, body, and soul. Because all these are naturally part of your extreme social needs, when you live your life underdeveloped, which are your needs for social competition and social supremacy. These are not normal social needs, but when you are underdeveloped, and when the environment is extremely austere, you switch to extreme social modes of life, when you seek social competition and social supremacy, ending up a tyrant or a dictator when you succeed.

And when you study society closely, everybody remains in these extreme social modes of life indefinitely, engaged in the continuous social competition and social supremacy, and therefore everybody may end up a tyrant and a dictator in this world. Creating all these social classes and social hierarchies, since when you seek competition and supremacy in this world, you are ready to serve, even in the most obedient manner, until you succeed and become a tyrant yourself.

Which is the case with everyone in this world, since as you study this world closely, everybody lives life at the zero addicted developmental level and at the first servitude consensual developmental level, with some at the second animal intuitive level, and with very few at the third intelligent human level.

Your social needs are sent to you directly by your social primal subconscious intelligence, since even your social needs are innate in you. Your specific extreme social need for social competition is meant to take you to the top of your entourage, to the top of your family and community, to the top of your nation, and then to the top of this world. And since this world counts in billions and since it is not only you receiving extreme social needs continuously throughout life, you do not actually get to make it to the top of this world, since others do. They dispose of you at will or they take control over you, and once they do so, you switch instantly to other modes of life, survival modes of life, when you are ready to do anything and serve anyone in any manner, only to be able to survive and to suffer less. And this is how you live your life, sometimes in extreme social modes of life and other times in survival modes of life, or even in intelligent modes of life.

Yet there is more involved here than you and them. From the perspective of your souls, this world may be just perfect as it is today, with all conflicts already present in order to stimulate you to give your best in this world, all meant to develop you the most. Because from the higher perspective of all souls, this world resembles more to an extraordinary resort or cruise ship, where most of the people are there to enjoy life. Other people are there to manage the entire enterprise, while a last category of people is there to serve and tend to all matters.

Who exactly are you now and who are they? Yet according to your souls, you might actually be among those benefiting from this entire enterprise significantly, with all social actors present there as simple personnel, tending to everything necessary. And this is the model of society now according to all higher beings: the clients, the management, and the workers. Or more precisely, the Elite are the management of the cruise ship, conducting the entire voyage. The Brotherhood are the personnel, providing the entire work throughout the cruise ship, while the Masses are the clients, those enjoying everything. What exactly is there to enjoy in life as Masses? But what exactly is there to enjoy in all videogames as the main character? Everything, while the Masses have it better than everybody else. Yet as you already notice, the current human civilization is not exactly an action videogame, but it is a highly intense one, already gone wrong, with most of the players already lost and disabled, incapable to remember who they really are.

Whatever the case is, you as an intelligence or living being of any reality have your rights since birth according to your status, and these rights are respected by anyone around, including all authorities of Earth, through the higher laws. The higher laws state simply that you have your rights to fulfill your needs continuously throughout life. Religions take this statement further to break it down into needs and how you have the right to fulfill them. Not only this, but the higher laws make a discrimination in the wider world, making more superior all intelligences and living beings that are more developed, and making more inferior the intelligences and living beings less developed. And this is how you have rights and status over the cows and chickens, and you may kill them and eat them, if you do so in order to fulfill your needs. Yet you cannot harass unnecessarily anything and anyone. Or if the higher laws are ever neglected, then any higher intelligence may intervene to restore them. And this is exactly how humans cannot be harassed and exploited unnecessarily, not by the Brotherhood, not by the Elite, neither by the invisible kingdom, nor by the higher beings that the Brotherhood, the Elite, and the invisible kingdom venerate.

What do they do? As you may notice, people always harm, harass, and exploit themselves in this world, with the invisible kingdom and the Elite never interfering. It only happens that everything ends up on their behalf. What they do, they have invented a specific manner to exploit and harm humans, by giving them a new, artificial identity, which is your artificial corporation name and identity, which is your name written in uppercase letters. That is a first level being or corporation, it has a first level status, and therefore it has first level rights, which is the bottom level. While your own natural living, intelligent identity has a third level status, and therefore third level rights. And you happen to reject your natural identity, status, and rights every time you carry and show around your licenses, permits, and ID cards, since they hold your name in uppercase letters and they belong to your consensual corporation, and not to you, not to your outer self, nor to your inner self. Because as a living, intelligent human being, you have the rights to fulfill your needs, all your needs, in any manner you can, without licenses, registrations, certificates, and permits. And this is why you are never employed, since no one may exploit you in any manner, but they employ your corporation instead. Which is the same, only that you have to identify yourself as a corporation every time you apply for work, every time you work, and every time you are paid.

Yet you do not have to live your life as a consensual corporation, or at least not if you do not desire it. As you notice, you have a multitude of natural selves in a multitude of realities higher and lower, including this world, and you may live freely in any of them, unless you choose otherwise. Would you ever choose otherwise? Would you ever choose slavery and divination, instead of freedom? Many people already choose so, even from here, from this world, yet you might choose slavery and servitude even as a soul, in order to be able to experience everything, including slavery, harassment, and exploitation.

But why having to live your life in other realities altogether? Why not living your life in your own reality, casually, since there is where you are born? Why having to descend into worlds as this one? Why do you use a phone when you can meet in person? Why do you surf the Net when you can read books? Why do you play videogames when you can do the same in real life? It is the same with living your life in lower, virtual worlds, since they keep you safe while they allow you to have ultimate experiences that would have endangered you and the entire world under normal, objective real circumstances.

Yet there is more involved here, since as we have seen, naturally created realities are used mostly throughout cognitive processes, and not only as entertainment. Because it does not matter what you do in life and how you do so, since you are still fulfilling your natural needs. Because for you, it may be simply playing a violent videogame because it is very fun, but for your security primal subconscious intelligence, you are actually fulfilling its major need of developing your mind and body in order to be able to cope with any undesired threatening circumstance, as a future war or any physical confrontation. Because as you see on TV in all news and movies, this world is highly violent out there, and therefore you always have to be ready to confront this world physically at all time, or this is the case according to your subconscious mind. Because your subconscious cannot distinguish between fiction, imagination, and reality.

And it is the same with your souls, because for them, it may be fun to live their life in our world alongside all their loved ones, while fulfilling their needs, as their needs for learning and understanding the wider world, developing consistently. Because as you see, the wider world consists of good and bad entities alike, and souls may fall into servitude and exploitation easily if they are not careful and developed enough to identify it, and now this is exactly as they train. And it is more complex, since there are even higher beings eager to help you as a soul to develop continuously throughout your multidimensional life. And this is how you end up in worlds as this one, which seems to be more intensive, because otherwise, you might have taken it easily, while living throughout calmer, forgiving worlds.

Or this is the case when you relax and entertain yourself, because if you achieve to understand your true meaning in life and in the wider world, then you do everything meant to fulfill your meaning, as this tends to have a cognitive nature. Because our world is not actually as a highly defined videogame as many souls portray it, but it has a cognitive meaning, and you as a mind, body, and soul should follow your cognitive meaning and do your part.

And this relates to the flat or spherical Earth, because if everybody followed their meaning, then this world was more consistent everywhere, lacking the dualities that we see all round. Yet as we notice, higher beings always become involved and stop you from following your natural meaning in life and in this world. And without being able to understand this entire social and living model of this world, you cannot do anything of higher importance. And since genuinely developed humans already have a natural need to learn everything necessary to fulfill their intelligent human needs, humans were already supposed to be able to identify all problems that they encounter throughout the fulfillment of their meaning in life and in this world.

Yet many people are not even aware that they have a natural meaning in life and in this world at their intelligent human level, spending their lives in addictions, servitude, and animal fulfillment. And even the Brotherhood, the Elite, and the invisible kingdom do not fulfill their intelligent human needs and meaning in life and in this world, but they also spend their lives in servitude and animal needs of all kind. Otherwise, they made this world a better place. And this is how you find genuinely developed humans only within the Masses, and not even there, because the Masses live lives in addiction and entertainment at the zero level, when they are not exploited at work at the first level. And again, these are not the third intelligent human level.

Well then, is the Earth flat or spherical? The Earth of our world is made in the correspondent image of all worlds above. And whatever the souls want, now this is the case here in this world. While as we notice, the souls even change everything drastically from one age of Earth to another. However, as you study our Earth and our world closely, you notice how everything is made in the image of a round Earth, or round space object, with all laws of physics applying to a round planet or space object. However, as you notice, this world is truncated in many instances, while currently, it seems that it does not include anything beyond the lower orbit of Earth. Everything else is there by default, as tides, day and night, light and heat coming from the Sun, and lights coming from all stars and planets, but all these do not have to be there objectively, since this is normally the case with all created correspondent realities. And if our world is as large as our Earth, then the Earth itself is actually a larger surface, where everybody lives their lives. Similar to all videogames, since all videogames are flat surfaces, regardless if they try to resemble a round world.

But now, when you study science and the entire current consensual society, you notice how some facts are hidden more drastically than others. Hallucinogens are banned the most, since they allow you to perceive the wider world, as slightly as they do. And then they hide the souls and higher worlds in every manner, probably to hide the created nature of this world. And then they hide the actual size of this world, not exceeding the lower orbit of Earth. There is no space out there, with all space missions fake. And then they hide the actual Consensual Matrix spanning the wider world, as it engulfs Earth and its higher worlds, with all humans and souls included. And then they hide the necessary continuous human development, enforcing instead sickness, addictions, and servitude. And then they hide the actual meaning that humans and souls have in life and in this world, leaving them doing everything else throughout life, but not fulfilling the actual meaning. Yet they do not actually hide these from humans, here in this world, but they hide it from souls and from their souls, here and in all their higher worlds.

While as you study the entire hidden human knowledge, you notice how nobody would live life underdeveloped anymore, throughout addictions and consensual servitude, since these are meaningless at the third, intelligent human level. And without servitude, the Elite, the Brotherhood, the invisible kingdom, and the dictatorships of the East are not possible anymore.

And now it seems to be even more tedious to model Earth and this world in several different manners, as it is in all higher realities, as it was supposed to be here in our world, as it ended up being here in our world, and then as it is modeled in all inner replicas of this world of all the people of Earth. Since all these models of Earth end up superimposed, to create the Earth that we know today. And as long as the people of Earth remain unaware of all these, this current model of Earth may remain stable, even as it is full of unseen dualities. Or it may be hijacked, to become anything anyone may desire, if the people of Earth remain ignorant of all these. And now, in order to find out more about all these assumptions, we have to use all information found so far, in order to create the model for Earth and for this entire world.

7 OUR EARTH, AND THE ENTIRE PLANE OF EXISTENCE

What you might not consider now is that you can have a flat Earth not only in the exact form that it is today, in perfect concordance with the azimuthal equidistant projection of the spherical Earth model that you are familiar with, but Earth may be flat in many other manners. And it might be only a coincidence that Earth is flat only in this manner. The official map of Earth today is this azimuthal equidistant projection, which may be the actual shape of Earth as seen from high above, and therefore the actual Earth.

We have gathered all the necessary information throughout this book. We have tried to gather everything available regarding the flat Earth from all domains and from all perspectives, with more information created on our own from additional data. And now we are ready to start the actual mental model, and if this is successful, then it becomes the actual model for the flat or spherical Earth.

Note that you still have to validate most of this data on your own, in order to keep out all possible beliefs. It is true that you can find several videos on the Internet showing exactly from how far away boats are seen through powerful binoculars right from under the curvature of Earth, but you have to perform this experiment yourself, repeatedly, obtaining very similar results in all conditions, in order to rule out any phenomena of mirage. Because it happens that major websites are owned by the invisible kingdom. Because this is why they keep the Internet centralized in major servers and not linked freely among all computers, in order to be able to own it, through enforced traffic. And you have to perform yourself all activity involved in gathering all your information for this model, just to make sure that it is based on accurate facts and not on beliefs, which is not a hard task. And then you have to make sure that your entire inner replica of this world is based on accurate facts and not on beliefs, in order to assure yourself an accurate, rigorous analytical reasoning. While this might be a tedious task, for you to go through everything that you know and understand, while verifying that everything is accurate and not stereotypical or believed.

Are we ready for the mental model now? We are actually going to have four mental models successively and even simultaneously, just because all readers have different minds, different reasoning, different beliefs enforcing their reasoning, and therefore we have to match all. We are still going to choose the most pertinent result, and therefore the most pertinent mental model. Because as already stated, in real life, it is not too easy, you do not have to perform only one mental model, since you are not always lucky to find your successful result or idea from the first trial. But you always have to modify your mental model, you have to find more information, you learn while you mental model, then you learn some more while you apply your results in practice. You have to modify your mental model again while you apply it in real world, and you learn some more. And this is how you obtain your results, this is how you learn from your results, and most importantly, this is how you learn from the entire procedure of mental modeling and physical activity, and this is exactly how you develop.

Our first of the four mental models that we undergo here is

the current model for the spherical Earth, as it is offered by science. Which you already know from school, since it is very simple. The Earth is a spherical space object.

Our second mental model is the model for the flat Earth set in place indefinitely by the Creator, which is the model offered by many religions today. We note from the start that there are many religions today controlled by the same social actors controlling this world, and this has been the case continuously throughout the ages. This may or may not be the case with your own religion, and in either case, you already know everything for this mental model, everything necessary to allow you to perform it right away.

The third mental model is the flat Earth model as it is offered implicitly by those controlling society. We are considering their model of this world here, and how they attempt to implement it in this world. In case that they ever attempt to do so, for the reasons mentioned throughout the book.

Our fourth mental model is the continuously developing model for Earth and for our world, as they are continuously changing while being created and co-created by all living beings and by all intelligences interacting with it throughout all existential clusters and lines of realities connected to Earth and to our world either directly or implicitly.

We start with the first mental model, which is the official model of Earth accepted today in this world, since all nations accept this scientific model, the spherical Earth. We already know that all the laws of physics apply to a spherical Earth, and most importantly, we know that they may manifest and take place only through a spherical Earth. The law of gravitation is based on a central force, the force of gravitation. While you can have this type of central forces only as they are made possible by round, spherical planets.

What science cannot offer is an accurate model for this world. Science claims that this world and the universe are the ultimate reality, and they appeared randomly in an explosion fourteen billion years ago. When you study this theory closely, you find it empirical, with the age of the universe determined randomly through trial and error, and made to be just a little older than anything that we may date around us. It is made through a simple, linear equation and a constant that you may change as you wish, to make the age of the universe older or younger. And this is not accurate knowledge, but erroneous, while this is not accurate research. And since all records from all space missions are fake, the entire model for this world as offered by science fails, including the spherical Earth. And this determines you to study the flat Earth model instead, because the current official model for this world fails.

Yet there is more to consider, since these same social actors controlling this world also control science along with all its scientific models, laws, and theories. And since they provide this world with inaccurate knowledge through a strong belief, now you cannot know what is true and what is false in this world according to science and according to these social actors controlling this world.

For example, if the spherical model of Earth is wrong and misleading, then the flat model has to be accurate, as they strongly believe. Yet now, by trying to implement the flat Earth theory in this world as we have seen, then maybe this is the inaccurate model, and not the spherical one. Yet what we must consider is that the Masses diminish continuously, while those controlling this world grow in numbers significantly. And this is exactly how all information related to them and to their beliefs becomes more available today, because this world becomes theirs. Yet this does not mean that the information that they believe in is accurate, since their beliefs and ideologies are very rigid and therefore very constraining, meant to be considered unconditionally.

We are ready now to start our first mental model, regarding the scientific, spherical model of Earth. Considering that all laws of physics are capable to offer an accurate basis for a spherical Earth, and considering that science fails to offer an accurate model for this world through the big bang theory, this does not mean that the Earth must be flat. Because maybe that science is incapable and unwilling to offer an accurate model for this world. We model now the entire world, as claimed by science and by all natural laws of this world, through spherical space objects and through space that exists outside the lower orbit of Earth, as far as the pictures that we see in the sky last. And it works. Space and this world spread further than the lower orbit of Earth, even further than all space objects that we may see at night with the most powerful telescopes. However, since all natural laws of this world and all images that we see in the sky at night apply to all upper realities to have created our world, and because our world is created in their image, all results found throughout our mental models apply to our higher realities directly, while they apply to our world only as far as our world exists. In fact, the main purpose of our world is cognitive in nature, and it is meant to do exactly this, to provide all information according to all realities above, and not according to this world. It is similar with all inner realities of your mind, since they are meant to offer you reasoning, solutions, and ideas according to this real world, and not according to their inner, truncated, specialized worlds. This is exactly why the higher worlds feed you premises, conditions, and implicit accurate knowledge about the entire outside world, in order for you to feed them successful solutions, knowledge, and ideas about the outside world, which you find and implement throughout your normal living.

And now our model starts making sense. Now we start understanding this world, and now we are capable to distinguish between what exists objectively in this world and what is placed here only as an image, only through assignments meant to match our upper realities, in order to determine us to reason, behave, and find successful solutions not according to this world, but according to all higher realities to have given the image of our world. And if you find statements throughout religious and spiritual records saying that this world was created in the image of the Creator, then we understand now that the entire image of this world does not have to be objectively real, but only assigned here, placed here for a specific purpose, a cognitive purpose, since life and behavior here in this world constitute reasoning, knowledge, solutions, and ideas throughout upper realities.

And now, you understand your meaning in life and in the wider world, which is to act naturally according to all these images, premises, conditions, and information, live your life normally and naturally according to them, reason normally, naturally, at your own developmental level, which is the intelligent human level, and this is how you fulfill your meaning in this world according to our Creator and according to the entire wider world.

The problem is that there are forces lower than the Creator, social forces and social actors from within this world interfering with your normal, natural meaning in this world and in the wider world. They interfere with your natural needs and your natural reasoning, this interference constrains you to live your life on lower developmental levels, and you become incapable to reason and behave at your intelligent human level. And this is exactly how you fail your Creator, since now, through all these imposed beliefs and stereotypes, you are incapable to reason and behave according to his expectations.

And then once you fail, you are doomed to be discarded by Life immediately. This happens with you, this happens with everybody, and this happens with the entire world. While you do the same with your mental models and entire reasoning, you discard them along with all faulty solutions and ideas that they have ever generated, to start again, hoping to succeed.

Therefore, to solve the problem of science now, this world lasts as far and as long according to all objective premises of this world, and not according to all premises, images, and conditions implemented here by our higher realities on their behalf. However, all higher worlds can use all reasoning and knowledge generated here in this world, since all premises and conditions here match their higher realities. Because all lower realities are specialized in particular lines of cognition, and therefore they are systematically truncated from the entire image of the higher worlds. And this is why you always have a multitude of parallel, specialized lower realities, together formed in the comprehensive image of the higher worlds.

This is the case with you and your cognitive system, this is the case with all videogame worlds, this is the case with this world, and this is the case with all created realities, either natural or artificial.

This is why your subconscious intelligence is divided in all primal subconscious intelligences, as your eating, reproductive, social, recovery, security, excretion, and respiratory intelligences, because these primal specialized intelligences truncate the outside world by specialized domains, by their specific specializations, and this is how they create their own inner replicas of this world now, in a specialized manner. Therefore, form the entire complexity and diversity of the outside world, your eating intelligence will create only an eating and digestive replica of the outside world, with all the food available in the outside world, where to find it, how to gather and store it, how to cook it and eat it and in what exact amounts, what nutrients it contains, how to digest it, how to distribute it to all cells of the body, how to store it within the body, and how to determine cells to eat it, when to eat it, and in what amounts. At the same time, your social primal subconscious intelligence creates its own specialized replica of this world in a parallel manner, containing all people of this world with their location and personality, and all social characteristics necessary for you to undergo a successful social existence. Your other primal subconscious intelligences do the same, and together with your own inner replica of this world, since you are the primal conscious intelligences, you truncate this entire world into parallel chunks of this world, which become self-sustained, viable inner realities, together processing cognitively the entire outside world, and together being able to reason and behave according to the entire outside world.

It becomes more important to understand how all inner intelligences of all your primal subconscious intelligences further truncate the already truncated specialized portions of the outside world, according to their inner further specializations that form your own upper perspective. And then, these inner intelligences use their own inner inner intelligences to do the same, together breaking down the entire outside world into smaller and smaller bits of knowledge and information, down to elementary accurate conceptual knowledge, and down to elementary algorithmic reasoning, as simple if-then and repeat-until thinking.

And now you may understand this world for what it really is, since it is an upward continuation of your entire cognitive system, it is an inner world as seen from above, an individual truncation and specialization of the entire upper reality. And this is exactly how everything objectively real present all around made it in this world, in this truncation of the higher world, while all images and assigned premises, conditions, and information are here only as replicas of the higher world, since they are not part of this truncation. And probably this is why this world may spread only as far as the surface of Earth, because everything else present in the upper reality is not necessary to be present here objectively in its image.

And this is exactly how Earth and our world may be accurate and objective all around us, probably up to the lower orbit of Earth, because the rest is only an illusion, placed there only to determine you to reason and behave just as you would do if you lived in the higher worlds. And it is done so only to make life, reasoning, and behaving on Earth as similar as possible to what takes place in the higher worlds.

How closely does this world resemble the higher worlds? We are exactly as close as Pac-Man resembled our world over half a century ago, or as close as Mario resembles this world today, as closely as your daydreams and social mental models try to match our world, and certainly as close as your own behavior can be while you undergo it as naturally as possible, free of the multitude of dogma, stereotypes, and ideologies present all around.

And yet, probably all stereotypes, dogma, and ideologies are placed all around on purpose, as they are always in the higher worlds. And therefore for you to have the chance to avoid them, since this is your meaning in this world. Should you actually join them just because they are placed there? Should you strive to avoid them just because they are placed there? I always encourage genuine natural reasoning, genuine natural behavior, and genuine natural development throughout life, according to all your genuine natural human needs and meanings that you receive throughout life, just because you receive all these directly from Life and from the Divine. Because if your own natural needs wanted you to behave on lower developmental levels as an animal, then Life would have brought a chicken or a cat in your place in this world, and not the extraordinary human being that you are. It is the same with dogmas, drugs, servitude, entertainment, and stereotypical thinking, because what need do Life and the Divine have from these in your life, if they downgrade your place, status, function, and meaning in this world? Because by using these, by living your life on lower developmental levels, you fail your genuine natural meaning in life and in the wider world, you fail your reasoning and development, you fail in life through your inaccurate decisions, you fail through your inaccurate results, you fail to understand this world including Earth, and therefore you are doomed to fail throughout all the future reasoning involving all true characteristics of Earth and of this world, since you fail understanding these in the first place. And this is exactly why all your higher needs converge now in this direction, pushing you to learn as much as you can about Earth and about the entire world, because all characteristics involving these are imperative throughout your future mental models.

So far, we are inclined to accept our fourth mental model to be accurate, the one involving the continuous creation of Earth, created by one higher being or by a multitude of co-creators resembling souls and even higher intelligences. In this manner, Earth is spherical in principle, as assigned by all higher worlds since it is probably spherical up there, just because all laws and information that we get here are consistent with a spherical planet and with an infinite world.

As stated, this world may be truncated and specialized, while lacking all unnecessary details for humans to perform their reasoning and to undergo their behavior according to all demands and conditions coming from up there. However, the third mental model is just as plausible so far, the one regarding the implementation of all the necessary premises meant to determine the entire world to accept that this world is flat, according to the artificial needs, ideologies, beliefs, and agendas of all social actors ruling this world.

Can the Elite, the Brotherhood, and the invisible kingdom be capable enough to fool the entire population of Earth into believing that Earth is flat and not spherical, for various reasons? Yet these social actors are capable to do everything on Earth, since they always do so anyway, mostly through you. And everything that you do yourself, you tend to believe to be true. The Brotherhood and the invisible kingdom can arrange and therefore order all flights in this world to take place not according to the spherical model of Earth, but according exactly to the azimuthal equidistant projection of Earth. And they can easily do so just because airplanes do not fly from point to point, because they are not allowed to fly from point to point due to a multitude of flight restrictions involving dangerous areas, military areas, and civilian areas. Consequently, what you have today is not free airspace for flight, but only free and assigned corridors of flight taking you around these restricted flying areas. And it is only a coincidence that all these corridors of flight match the azimuthal equidistant projection map of Earth and nothing else. And this may be done on purpose, in order to engage airplanes in the shortest flights to save time and fuel if Earth is flat, or it may be done on purpose to simulate a flat Earth out of a spherical one, for various reasons, and to make everybody wonder why all airplanes in this world behave as though the Earth is flat and not spherical. Because this will certainly make everybody believe that the Earth is flat and not spherical, just because airplanes fly in straight lines, or this is what people believe.

Even more, USGS has a significant budget to perform tasks that remain unknown, while USGS adopts this exact azimuthal equidistant projection map as the official map of Earth. Could this be the case just because the official belief of the invisible kingdom is still that Earth is flat, as they believed one thousand years ago when they were still an actual kingdom in Asia, just as all nations believed, and just as their current ideology still demands to believe today?

Yet not only airplanes have to travel across this world, but boats have to do so. Do boats and sailboats go through the northern hemisphere every time they have to sail from Australia to South America? No, not at all, but they just sail straight. You may do the same in your private yacht or sailboat if you choose to sail there. Even more, while sailing, you always use the spherical Earth model and you have to do so in a very precise manner, otherwise you run into islands and submerged rocks. Search the Internet, to find people sailing directly from Australia to South America or from Australia to southern Africa, only that they always choose to go north first, toward southern Asia, just because they sail in protected waters in this manner, going through better weather conditions. While it also takes them significantly less time, so we are back where we started, because these seaways going through the northern hemisphere may be shorter, because Earth may be flat again.

And do not expect this world to be spherical just because you manage to sail around this world, as science claims superficially today, since just because you manage to go around one or two continents placed on a flat surface, this does not mean that Earth is spherical. However, if you make it in time from Australia to South America, by covering the exact distance as it is stated according to the spherical Earth Model, and by not having to go through the northern hemisphere to get there, then Earth is spherical, and you have managed to prove it right. However, if it takes you over three times the distance to travel from Australia to South America, then you take the long way around assuming that it is the shortest way according to the spherical model of Earth, and therefore Earth is flat. Again.

To remain in our third mental model now, the Brotherhood and the invisible kingdom could have sabotaged all space missions and could have faked all space records on purpose, only to make everything to resemble their ideology, beliefs, religion, and agendas, stating that Earth is flat and that space and therefore this entire world spreads no further than the lower Earth orbit.

However, these social actors cannot fake your own discovery made at the beach, that you may see distant boats found already under the curvature of a spherical Earth. This might be a mirage or not, and this is what you have to find out yourself. As stated previously, mirages do not take place constantly on Earth, yet they may take place constantly on water, since the surface of water may be capable to refract and reflect light continuously, keeping it in this manner parallel to the curved surface of Earth. There is a very flat lakebed in South America, about one hundred kilometers wide, and when it rains, it fills up with water a few centimeters deep. The lack of winds and the very low depth make this lake extremely flat, while the lack of pollution and humidity in the air allows you to see all the way to the other side of the lake as though the surface of Earth is perfectly flat.

Yet as stated above, reflection and refraction alone are capable to maintain light parallel to the curved surface of Earth even indefinitely, making for a majestic view. Because wait until the entire lake dries again, and there are no reflection and refraction keeping light parallel to the curved surface of Earth, because you cannot see across the lake anymore, since Earth is spherical. Yet you might not be able to see across Earth anymore because the heat of the entire area creates thermal eddies in the air, altering the index of refraction of light everywhere above the lake bed, and this sends light in all directions, causing you to see nothing at all, not because the Earth surface is curved, but because light cannot pass in a straight line anymore above the hot, dried lake.

Again, you have to be there in order to perform all

experiments yourself, in order to be able to validate them. Because if you are capable to see normally across dozens of kilometers even above the dry lakebed, with no differences of temperature present at the surface of the lakebed, then the Earth is flat. And again, do not rely on the Internet in this research either, because they own the Internet and they censor everything posted online, allowing only misleading information.

Well then, we have to decide now which mental model is accurate, the third or the fourth one. And the answer is both, just because at the social level, you have these social actors acting according to their highly defined consensual needs, according to an agenda and ideology different than yours and different than the meanings of this entire world. And this agrees with our third mental model, as it even agrees with the true religion and spirituality, since these warn this world of these specific social actors meant to mislead you and your soul and to cause you to perish.

While from a higher existential perspective, it is possible to have an objective flat Earth, as it is the case in all videogame worlds, along with a perspective or assumed spherical Earth, resembling subjectively the shape that Earth and all space objects have within our higher worlds. And again, this agrees with our fourth mental model, while it also agrees with the true religion and spirituality in all instances. And even more, since all living beings and intelligences may co-create this world now and in the future, any strong belief addressing a flat Earth implemented systematically in the entire population may create a flat Earth, for the entire world to keep for some time or even indefinitely, according to the beliefs, agendas, and ideologies of all those controlling society.

You always have two choices, because you are either on the side of these social actors, knowingly or unknowingly, implementing everything that they demand, or you research everything independently, you reason for yourself, free of stereotypes, beliefs, dogma, and ideologies, you turn your back to all social actors intending to exploit you and the entire world, and you simply live your life as you should, at the intelligent human level.

The End

This book series continues with the next book, "The Human Environment." Here is a short synopsis:
You are born in this world believing everything to be accurate, reliable, and unique, just as you learn in school, and then as you grow up, this world shows itself in its true meaning. Parents have secrets at home and politicians lie and cheat throughout the media, while these might be the first inconsistencies that you notice in your perfect world. The rich become richer not exactly through their true abilities, but through their lies and illegalities, and this might be the first injustice that you ever notice in this world. Authorities replace authorities endlessly only to do the same, this harms you throughout life while authorities never admit it, yet you can see it well, everybody knows it, and this is your true, actual environment.

But can you pinpoint exactly what goes on in this world and what goes wrong in your environment? Because lies, inconsistencies, and illegalities are everywhere, and there is nothing significant enough offering you an accurate point of reference, an accurate perspective defining this altering environment, helping you understand this world as it is.

Because the knowledge found at the base of your understanding of this world is missing, as it is erroneous, or it is misleading on purpose, since this is how authorities strive to offer it to you. This alters your reasoning now, and consequently, it alters your inner, outer, and social behavior. Coincidently, your entire behavior is altered in this manner, determining you now to be part of these lies and illegalities yourself, directly or implicitly, determining you to create in this manner on your own this Consensual Matrix found all around, affecting and harming everybody, just as everybody's altered behavior creates the same Consensual Matrix affecting you and the entire world. Because the rich and the politicians never force you in any manner, but they only expect and accept your behavior to fall in this exact consensual pattern, coincidentally having all favorable outcomes turn in their favor. Because this enforced consensual environment that you create through your consensual behavior might be unfavorable and harmful to you and to those around, but it is certainly profitable for those controlling this world.

Because you must consider all environments when you study this world, as your natural, consensual, constructed, social, inner, cognitive, informational, higher, and spiritual human environments, forming your actual world. This defines your surroundings, it patterns your behavior and it gives meanings to your existence, it gives you a place and role in society and it challenges your existence, it constrains you to act and develop, to help and harm those around, it is there when you are born, and it always kills you in the end. And if you fail understanding your real and consensual environments for what they truly are, you end up missing a significant part of your existence, of what life truly represents to you and to everyone around, of what you should have always experienced, and of what you should have always been.

Throughout this book, we study the human environment in its entirety, natural, social, cognitive, spiritual, and consensual, since only through an accurate research of your entire environment you may find the accurate knowledge helping you structure and define your meaning and place in life and in this world.

ABOUT THE AUTHOR

Valentin Leonard Matcas, M.Ed., is a researcher, physicist, mathematician, educator, and an author of nonfiction and fiction books, including the entire "Human" book series. Valentin Leonard Matcas wrote the "Human" book series in the following order: "The Human Needs", "The Human Addictions," "The Hierarchy of Needs," "Stay in Shape, Lead a Healthy Life," "The Human Origins," "The Human Society," "The Human Conspiracy," "The Human Mind," "The Human Reality," "Astral Planes and Your Other Realities," "Life," "The Hierarchy of Intelligences," "The Human Intelligences," "The Human Thoughts," "Mental Models and Successful Ideas," "The Human Attitudes," "The Human Stereotypes," "The Human Ideology," "Modes of Life," "The Human Development," "Patterns of Development," "The Human Lifestyle," "Heal Yourself," "The Human Civilization," "The Human Religion and Spirituality," "The Human Rights," "Higher Laws," "Natural Laws of the Universe," "Existence," "The Human Condition", "Lifelines of Causality," "The Human Behavior," "Flat Earth," "The Human Environment," "The Human Meaning," "The Human Reasoning," "The Human Interconnectivity," "The Consensual Matrix," "The Matrix of Life," and "The Human Knowledge."
Valentin Leonard Matcas writes about terrestrial and alien civilizations, about life in the universe, the way it develops and intertwines across galaxies, about powerful beings as they control and reshape the universe, and about normal living human beings from Earth caught in this beautiful, wider, outstanding interconnectivity. Valentin Leonard Matcas creates a living, warmer universe in his books, teaming with life and vibrancy, on all levels of existence. Valentin Leonard Matcas also wrote "The Storyteller" book series, including "The Storyteller," "Starship Colonial," and "Unlimited," and "The Culling" book series, including "The Culling," "The Dream of the Dead," and "The Last Man on Earth."

When he does not work on his books, Valentin Leonard Matcas enjoys researching, hiking, swimming, kayaking, skiing, snowboarding, biking, reading, listening to music, and playing strategy videogames. You may discover all his books, videos, and articles.

www.ingramcontent.com/pod-product-compliance
Lightning Source LLC
Chambersburg PA
CBHW031422210526
45464CB00005B/2003